A Cubic Mile of Oil

A CUBIC MILE OF OIL

Realities and Options for Averting the Looming Global Energy Crisis

Hewitt D. Crane
Edwin M. Kinderman
Ripudaman Malhotra

OXFORD
UNIVERSITY PRESS
2010

OXFORD
UNIVERSITY PRESS

Oxford University Press, Inc., publishes works that further
Oxford University's objective of excellence
in research, scholarship, and education.

Oxford   New York
Auckland   Cape Town   Dar es Salaam   Hong Kong   Karachi
Kuala Lumpur   Madrid   Melbourne   Mexico City   Nairobi
New Delhi   Shanghai   Taipei   Toronto

With offices in
Argentina   Austria   Brazil   Chile   Czech Republic   France   Greece
Guatemala   Hungary   Italy   Japan   Poland   Portugal   Singapore
South Korea   Switzerland   Thailand   Turkey   Ukraine   Vietnam

Published by Oxford University Press, Inc.
198 Madison Avenue, New York, New York 10016

www.oup.com

Oxford is a registered trademark of Oxford University Press

Library of Congress Cataloging-in-Publication Data
Crane, Hewitt D.
A cubic mile of oil : the looming global energy crisis and options for averting it /
Hewitt D. Crane, Edwin M. Kinderman, Ripudaman Malhotra.
     p.   cm.
Includes bibliographical references and index.
ISBN 978-0-19-532554-6
1. Power resources.   2. Energy policy.   3. Petroleum reserves.   I. Kinderman,
Edwin Max, 1916–   II. Malhotra, Ripudaman.   III. Title.
TJ163.2.C73 2010
333.79—dc22   2009020237

9 8 7 6 5

Printed in the United States of America
on acid-free paper

Preface

History books and television series tend to give the impression that religions, emperors and kings, wars and treaties, invasions and rebellions, and pestilence and famine were the only truly important influences on the development of modern civilization. But it can also be argued that the availability of usable energy has been equally important, if not more so. Fire changed early man's life in many ways, as did the use of animals and simple mechanical devices such as the wheel, lever, and pulley. More complex mechanisms followed, including the steam and internal combustion engines and modern electrical and electronic technologies. All of these innovations have allowed humankind to continually expand its use of energy, and with it to improve living conditions and attain prosperity.

The oil shortages and price shocks of the 1970s made the industrialized countries conscious of their profligate use of energy, and individuals became aware of the benefits of conservation. Nations initiated schemes to encourage conservation in industry, residential, and commercial activities, as well as in transportation. These efforts resulted in substantial gains in energy efficiency. When we began the predecessor to this book more than a decade ago, the price of oil—the dominant source of energy—was declining in inflation-adjusted terms. Major oil producers, boasting of new oil and gas finds as well as improved extraction technologies, promised an assured supply of oil for 40 years.

With the decline in oil prices in the 1990s and the apparent "glut of oil," energy concerns dropped off of the radar screen for most individuals. Perusing the Web site of the U.S. Energy Information Agency for the historical prices of crude oil reveals that as recently as 2002, crude oil was selling on the spot market for less than $20 a barrel, essentially the same price, with adjustments for inflation, as that in the 1920s. Yet, the questions of where and how will we as a society obtain the energy to meet our future needs remained critical.

That fact is especially evident to those of us who engage in energy research. It takes many decades to effect changes in the pattern of energy use to bring a new technology from its first implementation to a useful scale. Even if we may have oil supplies for 40 years at our current rate of consumption, and even though we are likely to find additional sources, we will likely face the time when the projected increases in demand will outstrip our ability to produce oil at the necessary rate. At that time we will need additional resources. History has shown that obtaining energy from new sources takes a long time, particularly because the new sources often require developing a new infrastructure for their extraction, distribution, and use. That is why it is urgent to deal with the energy issue *now*: we can no longer ignore the impending energy crisis. A lack of awareness of the issues surrounding energy in the general media led us to title the predecessor to this book *Global Energy: The Ignored Crisis*.

Contrast the decline of oil prices in the 1990s with prices since 2005. Crude oil prices began a steady increase to $60 a barrel by early 2008, then soared to a high of $147 a barrel in July of that year, when just about every newspaper, magazine, or media report carried a story about energy. Although oil prices fell to around $40 a barrel by the end of 2008, energy is no longer an *ignored* topic. Energy-related greenhouse gas emissions and their effects on climate are constantly discussed and debated. World leaders propose and counterpropose new energy strategies for mitigating the impact of using coal and oil on climate. The public is being presented daily with options purported to be better for the environment. Yet, many of those options are no more than mere "greenwash." Some simply trade one problem for another, and still others do little more than make us feel good.

Renewable energy sources—which we refer to as income sources in this book—that could replace our inherited fossil fuels are being considered anew. However, can these income energy sources economically—if at all—replace the standard inherited energy sources? To make matters worse, the most abundant of the "standard" fuels are coal and heavy oils, which are by far the least desirable from an environmental point of view.

A large number of books on the subject of energy and climate change have recently appeared on the bookshelves. Only very few take a global view of energy production and consumption. Many of the books strongly advocate one technology or course of action and deride others. Some are filled with doomsday predictions resulting from global warming. Some even go on to specify 12-point plans to wean mankind off its addiction to oil and to live on renewables. On the other hand, there are books that claim that man-made global warming is the greatest fraud being perpetrated on the public, that solar energy will never scale, and that focusing on producing energy from renewable resources will have disastrous consequences for the economy and well-being of people. In this book, we have tried to stay clear both of cheer-leading slogans and of vitriolic diatribes.

Replacing one energy resource for another is a complex process, made even more complicated because it requires balancing the concerns of three E's: environment (including climate), economy, and the energy security of various nations. There is the real danger that in our effort to solve one problem, say, provide energy security, we can inadvertently create a worse problem, such as disrupting the food supply by diverting resources to produce fuel. We have made such mistakes in the past and must do everything to avoid them in the future. Exacerbating the situation is the immensity of the undertaking and the time and resources we will need to effect any substantial change.

This book's title, *A Cubic Mile of Oil*, was chosen to convey the sheer size of the challenge; a cubic mile of oil happens to be the current global annual rate of oil consumption. We have used it as a staple in this book and express the amounts of other sources of energy in equivalents of the energy contained in a cubic mile of oil (CMO). We believe that a readily comprehensible volume term—usable without exponential modifiers—helps laypersons and energy experts alike to better appreciate the scale of the problem we are facing. It is also the scale at which changes will need to happen. As we demonstrate in this book, it is very difficult to obtain energy at that scale from any of the alternative sources, so it will likely take a combination of many technologies to supply our energy needs in the future. Yet, in order to make any significant contribution, any technology will have to at least approach 1 CMO/yr, so in debating the merits of alternate energy sources, we have to consider what it would take to scale them to produce 1 CMO of energy per year and how their implementation at that level will affect us.

In 2007, we undertook a major revision of the book and updated all the statistics of energy production and consumption to reflect 2006 figures. While we were making these revisions, oil prices climbed from around $60 a barrel in early 2007 to $147 in July 2008, and then precipitously dropped to less than $40 a barrel by the end of 2008. These gyrations in oil price, and to some extent the prices of other fuels, reinforce the necessity to have a long-term energy policy spanning decades that is divorced from the prevailing price of oil.

The world currently uses about 3 CMO equivalents of energy annually, in oil and other sources combined, and by 2050 we will likely consume between 6 and 9 CMO/yr. Where we are going to get that energy and what it takes to produce even 1 CMO of energy are the questions we address in this book. The task is daunting enough even for established energy industries such as coal, oil, and nuclear to increase energy production by 1 CMO; it will require an even greater effort and commitment to increase energy production from sources like wind and solar that currently provide a miniscule portion of our energy.

As will be evident from the discussion that follows, replacing one energy source with another requires more than just a few years: it is a lengthy process requiring installation of a new energy-producing system, channels for its distribution, and diffusion in the market for the devices that can use it. The process can easily take decades. In the context of human history, a few decades is a very short time. However, when many experts believe we have only a few decades left before we must find alternative sources of energy to replace dwindling oil supplies and replace the current energy sources with those that do not emit greenhouse gases, a 50- to 60-year period is too long. For example, although coal use began at the beginning of the 18th century, consumption of coal did not surpass that of wood as the world's primary fuel until the late 19th century. Oil use began about 1870 but surpassed wood use only in the 1940s and coal only in the 1960s. The use of natural gas, which began about 1900, became about equal of that of coal only in 2000.

The relatively long time required for developing alternate sources of energy to replace the energy currently obtained from oil prompts us to label our current predicament a *crisis*. Oil supplies more than 90% of the energy for our transportation systems of cars, trucks, trains, and airplanes. While we are not "running out of oil," we are certainly finding it increasingly difficult to sustain production rates. Newer oil discoveries tend to be in places that are harder to produce from, and this crude oil often requires more processing. At the same time, the global demand for oil is increasing. Switching vehicles to run on alternate fuels will require a massive retooling, not something that we can achieve in a couple of decades. We could switch cars to run on electricity, but most of the electricity is produced from coal, a resource that emits more greenhouse gases and toxic pollutants than other fuels. Hence, coal is the least desirable fuel from the environmental perspective. There are also calls for reducing greenhouse gas emissions (mainly carbon dioxide) by 20% by 2020 and by 50% by 2050. While slogans may motivate us to engage, they alone are insufficient. This book addresses what it would really take to achieve these tough targets as well as some more modest ones.

Soon, the people of the world, through their leaders, must look far ahead and make major decisions about energy resource developments and energy uses. Increasing the efficiency of our energy use and employing conservation are two approaches that could have big impacts in a relatively short time frame. Implementing these practices offers our best hope for delaying the inevitable shortages, although doing so will involve expenditures of trillions of dollars and will influence activities and relationships in every part of the globe for decades to come. Increased efficiency and greater conservation will buy us the time to fully implement technical advances. Perhaps we will make some new discoveries with the potential to radically alter the situation; even ones that seem as remote as nuclear fusion—hot or cold—or finding a way to tap into "dark" energy,[1] or engineer a new way for capturing carbon dioxide from the atmosphere.

---

1. We do not completely dismiss the possibility of "miracles"—we recognize the limits of our knowledge. After all, physicists tell us that dark matter and dark energy, our *terra incognita*, make up 96% of the universe, and we know absolutely nothing about them. But, while miracles may happen, counting on them is not a strategy that we advocate.

As the factors cited in this book demonstrate, technological advances alone cannot deliver us from the energy crisis. With the spectacular improvements in the performance of computers that we have witnessed of late, society has come to believe that technology can solve any problem—"Just Google it" is a common refrain. Unfortunately, the laws governing energy are markedly different from those governing information. Finding, transporting, and using energy are not the same as finding, transporting, and using information. We hasten to add, though, that the ability to quickly find and widely disseminate information is critical to the rapid development of new technologies, and in that regard, the collaborative tools made available by Google and other Internet companies are a real boon to helping avert the looming energy crisis.

The great French marshal Lyautey once asked his gardener to plant
a tree. The gardener objected that the tree was slow growing and
would not reach maturity for one hundred years. The marshal replied,
"In that case, there is no time to lose; plant it this afternoon!"
—John F. Kennedy

Finally, despite the major influence that energy exerts on our lives, few truly understand its production and use. In preparing this book we have thus sought to explain energy in easy-to-understand terms to enable readers to enter more fully into the important debates and decisions that lie ahead. We believe it critical that energy debates take account of quantitative aspects—at least in an order-of-magnitude sense. To provide that information, we have mined data from numerous sources that address current energy use and future needs. We use the collected information in discussing different energy resources in terms of their total potential and also the effort required to develop them on a global scale.

On January 1, 2008, the *New York Times* carried a story about a Dutch company's plans to make use of hot asphalt on roads by laying down PVC pipes and flowing water through them to heat up homes. This sounds like a good plan, and it may indeed be one. We do not expect all the critical details to be present in a newspaper article. However, our hope is that after reading this book, such articles will immediately prompt questions like how much energy savings are we really talking about? The roads are going to be hot during the summer months, when the need for home heating is not as great. Roads often develop ruts and potholes; what is going to happen to the PVC pipes and the water flowing through them? Take another story. A company recently announced it has engineered a bug to convert waste into gasoline. That's an incredible feat and in no way are we discounting the accomplishment. However, if the bug relies on cellulosic or sugary waste, the question to be asking is whether there's enough waste. As we shall see in the chapters that follow, even in our highly wasteful society, the waste we produce contains very little usable energy—perhaps a few percent of the energy we use. If questions like these arise in the minds of our readers, we will feel that our book has served a key purpose.

There are no easy answers, which basically means that everyone will find some elements of every solution to be objectionable. There are trade-offs to be made at every step of the way. The democratic way to resolve such issues is through an open and informed debate based on facts without resorting to disdaining those with different viewpoints. Our hope is that this book provides the facts about the various competing technologies that are necessary for making informed choices and for stimulating a rational debate. We doubt that an immediate and international consensus will emerge out of such a debate but sincerely hope that sufficient numbers of people find common ground to work with and enlist ever larger numbers of people to join them in addressing this crucial issue. Our book is intended to provide the seeds for informed discussion and decision making as communities and nations align themselves to address the future.

For access to supplemental material, including a Q+A with the authors and full-color images of the exhibits, visit www.oup.com/us/CubicMile.

Acknowledgments

We thank Dr. Curt Carlson, CEO of SRI International, for his encouragement, enthusiastic support of this project, and spirited discussions in developing many of the ideas expressed in this book. He, and his predecessors, aided us in writing this book by making many of the resources of SRI available to us. We are indebted to many of our colleagues and clients with whom we have worked over the years and who helped shape our thoughts. In particular, we thank Roger Sherman for assistance with library research, Julie Kirkpatrick for ably typing the draft manuscript, Michael Smith for insightfully editing it, Scott Bramwell for his assistance with the graphic arts, and Rajan Narang for his help in the preparation of the final manuscript. We are grateful to Jeremy Lewis, Edward Sears, Brian Desmond, and Mary Kaufman of Oxford University Press during the various steps of publication of this book, and to Trish Watson for her thoughtful and thorough copyediting.

We also acknowledge the help and useful discussions with many colleagues and friends who offered valuable criticism and suggestions. At the risk of leaving some names out, we mention the following individuals, who were particularly helpful: Marcello Hoffman, Jack Goldberg, Charlie Rosen, David Huestis, Robert Wilson, and Michael Balma. Our special thanks to Donald McMillen and Marianne Asaro for numerous conversations and their critique of the manuscript, pointing out inconsistencies and suggesting ways to make the arguments cogent.

On a personal note, Hew thanks his wife, Sue, and his sons, particularly Doug, who provided encouragement at every step of the way. Ed acknowledges the moral and physical support from his many progeny, especially his son Albert, who provided help with several mathematical constructs. Ripudaman thanks his wife, Ellen, for supporting this project and not only giving him the time to work on the book but also serving as a sounding board, critiquing many of the earlier versions of the book, pointing to where the discussion was bogging down in details or was rife with jargon. Ripudaman also thanks the community of Hyde School in Woodstock, Connecticut. In a profound way, they influenced him to join Ed and Hew to complete their book.

Contents

PART III: THE PATH FORWARD

The Authors

The three coauthors of this book are, or have been, employees of SRI International, an independent nonprofit research institute in Menlo Park, California, that conducts client-sponsored research and development for government agencies, commercial businesses, foundations, and other organizations. While all three authors are interested in energy matters, they came to that interest from distinctly different backgrounds and from different divisions of SRI, and before writing this book had never worked together. It was their passion for the role of energy in our and our children's lives that brought them together to write this book.

Hewitt Crane (Ph.D., electronics, Stanford University, 1960) came from the world of electronics and computers, starting in the late 1940s with work on an early IBM vacuum tube and relay machine, followed by involvement with the first all-electronic computer at the Institute for Advanced Study in Princeton, and then the first computer for banking built at SRI (known then as Stanford Research Institute). His highly acclaimed works on digital-logic technologies, multiaperture magnetic devices, and a noninvasive eye-movement tracking instrument earned him a fellowship of the Institute of Electrical and Electronic Engineers. Since that time, his interests turned toward the fields of neurophysiology, human sensory systems, and more recently cosmology—none of which has the slightest connection with the world of energy. However, Hew did have two rather remote links. The first was an infatuation with large numbers, such as those involved in biological systems, and in finding ways to express them so that their true magnitude could be more easily visualized. That love was put to the test during the early 1970s while he, like many in the

United States, sat in long gas lines. "How much oil is used annually?" suddenly became a challenging question. When he found that the global number at that time was beginning to approach a scale of a trillion (1,000,000,000,000 or 10^{12}) gallons per year, he searched for ways to convey the true magnitude of that quantity. He found that a cubic mile contains almost exactly one trillion U.S. gallons; out of this coincidence evolved the "cubic mile of oil," or CMO unit, that is used throughout the book; for all practical purposes, 1 CMO/yr accurately the current annual global consumption of oil, and 3 CMO/yr represents the world's total annual energy use.

Hew put this fact aside until Iraq invaded Kuwait in 1990, when his interest in energy was rekindled and a host of new questions arose with a similar link to energy. How much oil is left? How many of the other fossil resources are being consumed, and how many are left? What is the status of alternative energy? What lies ahead?

Hew was appalled at energy waste in living and working environments and at the necessity of reliance on military force to assure the U.S. oil supply. He suggested a vigorous energy conservation program at SRI in the wake of the first Gulf War, and in short order SRI's annual $3.5 million electricity bill was reduced by more than $1 million, to say nothing of the energy conserved. In 1994, about a year into the program, he mentioned to SRI's president his interest in preparing a series of notes on energy to distribute to the staff to help maintain the conservation momentum, and that he intended to use the CMO unit as a simple way to answer some of the questions noted above. Little did he know at the time that he would join forces with another SRI researcher, Edwin Kinderman, who had worked in the energy field since the early 1940s, had a great deal of first-hand knowledge of many aspects of the field, and could thus provide in-depth perspective and informed analysis. That is how this book was conceived. Hew was one of SRI's most prolific inventors and visionaries and was named an SRI Fellow in 1985. He authored *The New Social Marketplace: Notes on Effecting Social Change in America's Third Century* and coauthored *Digital Magnetic Logic* with David Bennion and David Nitzan. He retired from SRI in 2001 and passed away in 2008 before he could see this book in print.

Edwin Kinderman (Ph.D., physical chemistry, University of Notre Dame, 1941) comes from the world of physics and chemistry. He began work at SRI in 1956, just two weeks before Hew Crane. By then, Ed already had an extensive background in energy, starting with isotope separation and uranium processing research at the Radiation Laboratory of the University of California in the early 1940s, followed by seven years of research related to nuclear reactor operation and nuclear fuel separation at the Richland, Washington, plutonium production plant. Following that, he was engaged in a diverse set of projects on special nuclear material control, chemical processing, and reactor operations at the General Electric Company. At SRI, while directing the diverse energy-related, laboratory-based programs of the Applied Physics Laboratory, he personally conducted research dealing with applications of ionizing radiation and reactor safety issues, as well as nuclear materials control, terrorist and rogue nation nuclear threats, and proliferation issues. The groups under his direction dealt with subjects as diverse as lasers, energy exchange between atoms, radiation emitted by nuclear explosions, superconductor-based measuring devices, magnetic train levitation, and the transfer of government-developed technology to industry.

In the early 1970s, Ed changed his focus to studies of energy end uses and conservation, and by 1980 he was fully occupied with technoeconomic and market analyses for government and industrial clients concerned with the broader spectrum of alternative energy sources. While much of his work dealt with industry confidential and government classified matters, he has published a number of research articles in peer-reviewed journals and contributed chapters to two books on nuclear proliferation. He also presented two invited papers on proliferation issues at the first International Conference on Nuclear Material Safeguards in Vienna in 1965. Although formally retired since 1994, Ed is still actively investigating alternative energy resources and technologies and environmental control policies, while devoting time to the analysis of the broader energy issues addressed in this book.

Ripudaman Malhotra (Ph.D., chemistry, University of Southern California, 1979) is associate director of the Chemical Science and Technology Laboratory in SRI's Physical Sciences Division. He is an organic chemist, and during his tenure at SRI he has worked extensively, though not exclusively, in the area of energy. Most of his studies have focused on the chemistry of processing fossil fuels. His detailed mechanistic studies of these systems have resulted in innovative processes that achieve desired product selectivity and increased efficiency. As someone deeply engaged in energy research, he was acutely aware of the looming energy crisis, which was being exacerbated by the potential of global climate change. He broadened his research interests into studying alternate resources such as biomass and application of biotechnology in the areas of energy, chemicals, and the environment. He recognized the value of the CMO unit and had been citing the unpublished work of Crane and Kinderman for about a decade. In 2005 he joined Hew and Ed to help finish their book. Among his published works are more than 90 papers in archival literature, coauthorship of a book on nitration, editorship of a book on combinatorial materials development, and coeditorship of a book on advanced materials. He is also the section editor for a multi-volume Encyclopedia of Sustainability Science and Technology, responsible for the chapters on the production, uses, and environmental impacts of fossil resources. The encyclopedia is scheduled for publication in 2011. He is an active member of the Petroleum and Fuels Chemistry divisions of the American Chemical Society. In 2005, he was named an SRI Fellow.

List of Exhibits

Some Common Energy Conversion Factors

To convert values expressed in CMOs to any of the units shown on the left, multiply the number of CMOs by the multiplier shown on the right. To convert values in those units into CMO, divide by the multiplier shown.

Unit of Measurement and its Abbreviation	Field of Use	Multiplier[a]
Barrel (bbl)	Oil	26.2×10^9
British thermal unit (Btu)	General	153×10^{15}
Calorie (cal)	General	38.7×10^{18}
Cubic meter (m³)	Oil[b]	4.35×10^9
Gallon (gal)	General	1.10×10^{12}
Joule (J)	General	162×10^{18}
	Electricity	
Kilowatt-hour (kWh)	• End-use energy	15.3×10^{12}
	• Primary energy equivalent	45.0×10^{12}
Quad	General	153
Standard cubic foot (scf)[c]	Natural gas	153×10^{12}
Terawatt-hour (TWh)	End-use electricity	15.3×10^3
Terawatt-year (TWy)	End-use electricity	1.75
	Primary energy equivalent	5.13
	Coal[d]	
Ton (t)	• Hard coal (12,000 Btu/lb)	5.10×10^9
	• Brown coal (6,000 Btu/lb)	9.56×10^9
	Coal[d]	
Tonne (metric ton; te)	• Hard coal (12,000 Btu/lb)	4.64×10^9
	• Brown coal (6,000 Btu/lb)	8.69×10^9
	Oil	3.84×10^9
Trillion cubic feet	Natural gas	153

[a]Legend for exponents: 10^9 = 1 billion (1,000,000,000); 10^{12} = 1 trillion (1,000,000,000,000); 10^{15} = 1 quadrillion (1,000,000,000,000,000); 10^{18} = 1 quintillion (1,000,000,000,000,000,000).
[b]Oil energy values vary somewhat with the quality of the oil. We have used 40.0 MBtu/tonne as the heating value of oil, which is the same as that used by BP (formerly British Petroleum).
[c]"Standard" refers to conditions of pressure (1 atmosphere) and temperature (60°F).
[d]Coal values are approximate because coals vary substantially in their heat content.

Symbols and Abbreviations

ABWR	Advanced Boiling Water Reactor
AC	alternating current
ANWR	Arctic National Wildlife Refuge in Alaska
bbl	barrel
bpd	barrels per day
Btu	British thermal unit
BWR	boiling water reactor
CAFE	Corporate Average Fuel Economy
cal	calorie (thermal)
Cal	food calorie (= 1 kilocalorie, or 1,000 calories)
CFL	compact fluorescent lamps
CO	carbon monoxide
CO_2	carbon dioxide
CSP	concentrating solar power
CTL	coal to liquids
DARPA	U.S. Defense Advanced Research Projects Agency
DC	direct current
DOE	U.S. Department of Energy
EIA	U.S. Energy Information Agency
EPA	U.S. Environmental Protection Agency
eV	electron volts
FAME	fatty acid methyl ester
FBR	fast breeder reactor
ft^3	cubic foot

gal	gallon
GDP	gross domestic product
GHG	greenhouse gas
GO	gallon of oil equivalent
GTL	gas to liquids
GW	gigawatt
GWP	gross world product
HDR	(geothermally) heated dry rock
HVAC	heating, ventilating, and air conditioning
IAEA	International Atomic Energy Agency
IEA	International Energy Agency
IGCC	integrated gasification combined-cycle
IPCC	Intergovernmental Panel on Climate Change
J	joule
kcal	kilocalorie = 1,000 thermal calories = 1 food calorie
kWh	kilowatt-hour
LED	light-emitting diode
lm	lumen
LPG	liquefied petroleum gas
LWR	light water reactor
m^3	cubic meter
Mbpd	million barrels per day
MBtu	million Btu
mi^3	cubic mile
MJ	megajoule
mpg	miles per gallon
MW	megawatt
MW_e	megawatt electric; specifies electricity generation capacity
MWh	megawatt-hour
NEA	Nuclear Energy Agency
NHTSA	National Highway Traffic Safety Administration
NiMH	nickel-metal hydride
NRC	U.S. Nuclear Regulatory Commission
OECD	Organisation for Economic Co-operation and Development
OPEC	Organization of the Petroleum Exporting Countries
OTEC	ocean thermal electric conversion
ppm	parts per million
psi	pounds per square inch
PV	photovoltaic
PWR	pressurized water reactor
quad	10^{15} Btu
R&D	research and development
RAR	reasonably assured resources
scf	standard cubic feet
SNG	substitute natural gas

SUV	sport utility vehicle
SVO	straight vegetable oil
t	ton (0.91 tonnes)
te	tonne (metric ton)
TWh	terawatt-hour
TWyr	terawatt-year
UNSCEAR	U.N. Scientific Committee on the Effects of Atomic Radiation
USGS	U.S. Geological Survey
USMMS	U.S. Minerals Management Service
WEC	World Energy Council
Wh	watt-hours
WIPP	Waste Isolation Pilot Plant
WNA	World Nuclear Association
yr	year

A Cubic Mile of Oil

1

Introduction

Energy is central to our existence and our way of life. We use it in virtually all aspects of life: manufacturing the myriads of goods that we have come to depend on, growing our food, transporting goods and people, controlling our environment, communicating with one another, entertaining ourselves, and the list goes on. By and large, the standard of living of a society is directly linked to its energy consumption. Indeed, today's technological society can be described, quite literally, as "turning oil into everything else that we eat or use." Energy use is so pervasive that we often fail to recognize its role and are only reminded of our dependence on it when for some reason or another there is a shortage. To be sure, such reminders have occurred and will occur every so often. However, the shortages we have overcome thus far are very minor compared with what may lie ahead. With ever increasing numbers of people and nations striving to improve their standard of living, the demand for energy is soaring. At the same time, traditional sources of energy are being depleted, and even their current level of use poses a serious threat of global climate change. How are we going to provide the vast amounts of energy that we will need or desire in the future? That is the central question that this book addresses.

Effective resolution of any major societal issue requires easy access to reliable and relevant information. When an issue is not only complex and multifaceted but also essential to maintaining the very fabric of the society, a lack of comprehensible information makes the public's role and government leadership less effective, and appropriate solutions become more difficult to implement. This is the situation today with global energy—arguably the world's largest industry and one central to all our lives. On the one hand, the public is generally unaware that a pervasive, long-term problem exists and that the world is facing a complex and potentially perilous, perhaps even revolutionary, future. On the other hand, experts and representatives of different segments of the vast energy enterprise have different and frequently conflicting solutions for solving the energy dilemma, and vastly differing motivations for deemphasizing one aspect or another of the problem. Moreover, commentaries in the mainstream media are often highly polarized and politicized, leaving readers or listeners bewildered about whom or what to believe.

The complexity of the energy world, the urgent need to take action, and the difficulties involved in understanding both the current and potential needs for energy and the technologies (and their potential for development) led to the writing of this book. We have sought to provide basic information about the technologies and the technical advances needed to produce future energy supplies. We believe it is important that this information be available not just to the scientists and engineers who specialize in energy issues but to people from all walks of life: a student of literature concerned about the human condition; the city planner deciding how best to provide a greener environment with limited resources; the senator (and her staffers) debating the policy options for securing our energy supplies; the venture capitalist looking to invest in clean technologies. As already mentioned, making substantial changes to the global energy supply presents a challenge of gigantic proportions, and it will affect the lives of all people—some more so than others. It will require many creative and innovative ways of overcoming those challenges. It will require fresh thinking and input from people from different backgrounds.

Scientists and engineers may come up with some solutions, but it will require business leaders and entrepreneurs to bring those ideas and solutions into the market and into the hands of the customers. There is also a critical role for government. Many of the products and solutions probably do not have ready access to the energy market, and in some instances those markets may not even exist today. By exercising certain policies, governments can create those markets. Another important role for the government is setting a level of tax on the different sources of energy that is most appropriate for its circumstance. Keeping fuel prices low may help the local economy, but it could also lead to wasteful practices. Government representatives, in turn, respond best to public demand, and because any policy actions (or inactions) are likely to affect the lives and livelihoods of millions of people, the public at large must enter into an informed debate. Accordingly, in writing this book we have attempted to present the information needed for engaging in an informed debate accessible to a broad readership by explaining basic elements of different technologies. Although specialists may find our explanations overly simple, we have sought to provide enough depth and detail so all readers can appreciate the pros and cons associated with the different choices about future sources of energy. After all, the decisions about energy choices affect the public at large, and therefore the public has to engage in the deliberations.

A Veritable Tower of Babel

As we began preparing this book, we found it frustrating that each source of energy is described with a different set of units. Production and use of coal are expressed in tons (or tonnes [metric tons]); oil in tonnes, barrels, or gallons; natural gas in standard cubic feet (scf), and so on. Although many books and articles discuss energy sources in terms of their energy content in a common unit, whether in British thermal units (Btu), joules, calories, or watt-hours (Wh), each of these units represents a relatively small amount of energy, and mind-numbing multipliers such as billions, trillions, or even quadrillions must be applied when discussing their use in the global context. A veritable tower of Babel results. Besides being unable to relate to those units (how much is a Btu, a joule, or a quad?), we found it very hard to keep those large numbers straight in our heads.

The use of power units such as gigawatts (GW) or terawatts (TW) in the context of energy is another source of confusion. Power is the *rate* at which energy is produced or consumed, although in common parlance power is also used to refer to electricity. It is true that we can describe annual global energy consumption in power units and therefore state the challenge for the future in terms of requiring, say, an additional 20 TW of power. However, production capacity alone does not tell the whole story. The energy produced in one year by a 1-GW coal plant is much different from a seemingly equivalent 1-GW wind installation, because a coal plant typically operates at its rated capacity about 85% of the time, while a wind farm installation does so for only around 25–30% of the time. The net result is that it takes about three times as much installed power capacity for a wind power system as for a coal power system to produce the same amount of electrical energy. Likewise, the capacity factor for solar power systems is generally less than 20%, so if we decide to go exclusively solar, our need for the additional 20 TW-year of energy will translate into a need to install about 100 TW of capacity.

In discussions of global energy and resources with our friends and colleagues, we found that many of them shared our frustration with all of the different units being used to describe energy. What we needed was a large unit of energy that could be visualized and would also evoke a visceral reaction. It is perhaps ironic that while we bemoaned the fact that we already had too many units for energy, we have ended up introducing yet another.

Cubic Mile Oil Equivalent

We turned to a unit that one of the authors, Hew Crane,[1] had devised as he sat in the long gasoline lines that typified the energy crisis of 1974. He had heard

---

1. We sadly mourn the passing of our friend and coauthor, Hew Crane. It is unfortunate that he did not get to hold the finished book in his hand.

that the world was using oil at the rate of about 23,000 gallons a second and began wondering how much would it be in a year. A few multiplications later, he calculated it to be approaching a trillion gallons; 724 billion gallons, to be more precise. Hew was always fascinated by large numbers, whether they be the number of cells in our central nervous system or the number of ants on Earth. He also had a passion for devising ways to communicate their magnitude. Unable to picture the trillion gallons, he began calculating how large a pool could contain that amount. A few more arithmetic steps later, he realized that if those 724 billion gallons of oil were to be poured into a pool a mile wide, a mile across, and a mile deep, the pool would be about three-quarters full.

That was the genesis of a cubic mile of oil, or CMO, as our new unit and the staple measure used in this book. Incidentally, a CMO is the amount of oil the world used in 2000; in 2006, use had increased to 1.06 CMO. A cubic mile is the same as 1.10 trillion gallons or 26.2 billion barrels. However, the unit allows us to dispense with modifiers such as billions and trillions when talking about global oil and other major resources. Whereas in the United States we take billion to mean a thousand million (10^9), and trillion to mean a million million (10^{12}), such is not the case throughout the world. In some parts of Europe, a billion is a million million. And thus, avoiding the use of these terms removes one source of confusion. We have often encountered articles in print that mistakenly refer to a billion when they meant a million, and so on. A case in point is an editorial describing that the potential savings in oil from increasing the efficiency of cars and trucks in the United States from a fleetwide average of 25 mpg (miles per gallon) to 35 mpg would amount to a *billion barrels per day*.[2] Now, the total U.S. oil consumption is only 20 million barrels a day, so immediately we realize that something is amiss: how could you save a billion barrels when the total consumption is only 20 million? An increase in fleetwide efficiency from 25 to 35 mpg could easily lead to savings of a *million barrels a day*, and perhaps even 3 million barrels a day, for a total of a *billion* barrels a *year*. So perhaps the editorial was referring to annual savings. The point is not to pick on this piece or its writer, but to illustrate how easy it is to get lost when the numbers are large and hard to comprehend.

A volumetric unit also makes it possible for us to form a mental picture: a cubic mile is a pool a mile wide, a mile long, and mile deep! If that is hard to comprehend, try picturing a large sports arena, say, 700 feet in diameter and 250 feet high (about 25 stories). The volume of this cylindrical arena would correspond to 96.2 million cubic feet, and it would take about 1,500 of these arenas to equal the volume expressed in a cubic mile. The Bird's Nest Stadium, where the opening ceremony of the 2008 Olympics in Beijing was held, was the world's largest sports facility at the time, with an official volume of about 4.9 million cubic meters. A cubic mile of oil (or any liquid, for that matter)

2. "Time, finally, for real fuel economy," *New York Times*, August 2, 2008.

could fill the Bird's Nest Stadium 850 times. A CMO is a unit of energy, and when we refer to CMO, we are referring to the energy content of a cubic mile of oil: the thermal energy released during its combustion.

Most data in this book are given in regular English units because the book is primarily addressed to a nontechnical U.S. audience, although in many instances we have also provided their metric equivalents. A cubic kilometer of oil (CKO) measure might be more familiar for the international community. There are 4.17 CKO in 1 CMO, but for our purpose, even a simple multiplication by four would suffice to arrive at the CKO equivalent of the values quoted here in CMO units.

The thermal energy released during the combustion of a cubic mile of oil is 153 quads (153×10^{15} Btu), and we can use the thermal energy content of other fuels such as coal and natural gas as the basis for expressing their amounts in equivalent volumes of oil. In 2006, in round numbers, the world consumed about 0.8 CMO of coal and 0.6 CMO of natural gas, in addition to the 1.1 CMO of oil. Combined, the world used 2.5 CMO of energy from these three fossil resources—coal, oil, and natural gas.

Describing the energy derived from nuclear and hydropower plants is less straightforward. These plants produce electricity directly, which we use in a large variety of ways at home, in the office, and in industry. We could compare these two power sources on the basis of the energy of the electricity they produce by using known conversion factors (1 kWh = 3,414 Btu), but doing so would ignore a salient fact: electricity is generally *used* with very high efficiency but is commonly *produced* with only modest thermal efficiency. When electricity is produced from coal, oil, or natural gas sources, about two-thirds of the energy in the fuel is lost in the process. Strictly speaking energy is conserved, with the extra energy being dissipated into the environment as heat from which we cannot derive any useful work. Currently, an average thermal power plant produces 1 kWh of electrical energy from 10,000 Btu of fuel.

Accordingly, in expressing the output of hydroelectric, nuclear, photovoltaic, geothermal, and wind power systems in CMO units, we have used 1 kWh as if it *equaled* 10,000 Btu to arrive at their thermal energy equivalent. With this approach, 1 CMO is equivalent to 15.3×10^{12} kWh, as opposed to the 44.8×10^{12} kWh that would be obtained by simply considering energy equivalence. We note that now both the U.S. Energy Information Agency in its various documents and the petroleum company BP in the *BP Statistical Review of Energy* make an adjustment, albeit a somewhat different one, for expressing electrical output from fossil energy. Incidentally, the global generation of electric power in 2006 was 19 trillion kWh, or 1.2 CMO.

With this consideration for the generation efficiency of electricity in mind, we calculate that the world uses somewhat less than 0.2 CMO each of hydropower and nuclear power. Biomass provides around 0.15 CMO, primarily from burning wood. Contributions from biofuels (i.e., fuels such as biodiesel and ethanol made from plants), wind, photovoltaic, and solar thermal currently barely register on the CMO scale. The nearly 2 billion gallons of biodiesel and 15 billion gallons of ethanol globally produced in 2008 together amount to 0.01 CMO. Likewise the total energy generated from the installed 16 GW of PV systems and 120 GW of wind turbines amount to 0.02 CMO. Adding the contributions of nuclear, hydro, and biomass sources to the 2.5 CMO from the fossil resources results in grand total global energy consumption equivalent to more than 2.9 CMO, or say 3 CMO.

To be sure, a CMO represents a very large amount of energy and is most appropriate when discussing global and national energy use, but it is not a convenient unit for describing energy use by individuals (i.e., per capita use). Where per capita energy consumption is discussed, we have used gallons of oil equivalent (GO) units. Again, although there are 3.78 liters in a gallon, we can approximate the metric equivalent of GO, liters of oil equivalent (LO), by simply multiplying the GO values by four, which happens to be the same factor that gives us a useful conversion from CMO to CKO!

Even though a CMO is not a quantity that we can put our arms around, we hope that the simplicity of the CMO and GO concepts will enable communication among the public, as well as among experts in the separate energy communities. We also hope that this more approachable unit will facilitate the development of easy-to-read and understandable books and other media (including Web resources) dealing with energy facts and issues for all levels— from grade school through university.

Another factor that complicates the situation is that over the years we have heard many predictions of the world running out of oil in 30 to 40 years. Many of these "predictions" are based simply on the estimated volume of oil reserves available and the current rate of oil consumption, without consideration of future changes in either. The reserves/production ratio has stood at between 30 and 40 for more than 40 years. This ratio has been erroneously taken as a measure of how long our oil supplies will last. The assumption is in error because, on the one hand, new findings, new technologies, and an increase in the price, add to the reserve base, thereby making the oil last longer, and on the other hand, increased consumption rates shorten the period in which the reserves will be exhausted. In a way, the constancy of the reserves/consumption rate ratio suggesting a 30- to 40-year supply means that oil producers have little incentive to look for more oil, and that they reason they are better off making further investments elsewhere, say, in improved refining. Because of the erroneous predictions, the public has grown highly suspicious of experts making more predictions along those lines.

In fact, this book is not so much about *when* we will run out of oil, as it is about *whether or not* we will able to supply the world with energy at the rate that it is likely to demand. As we discuss in chapter 4, our future energy needs are based on several projections. Most of these projections forecast a substantial increase in global energy consumption resulting from economic development in the emerging economies with large populations. These projected increases are also on the order of one, or more, cubic miles of oil. Thus, by 2050, not only would we have to find alternative sources of energy to replace our current oil reserves, but we would also have to find additional sources of energy to meet the expected increased demand.

Global Warming: The Elephant in the Room

In psychotherapy, the elephant in the room refers to the hugely important topic that no one wants to talk about. In both the scientific community and in the general public, there is a great deal of discussion about global warming— much of it inaccurate or hyperbolic—and we are not going to take a position on it in this book. We will tiptoe around this elephant.

Coal and unconventional petroleum sources, which are plentiful, could be exploited to meet the increasing energy demand. But can we continue to use the fossil energy resources that produce billions of tons of CO_2 a year with impunity? Are the emissions from our use of fossil fuels causing the global warming? And, if the answer to these two questions is yes, what are the consequences of the temperature increase? In 2007, the Intergovernmental Panel on Climate Change (IPCC) published its Fourth Assessment Report on the issue[3] and concluded that the warming of the globe due to greenhouse gases (GHG) was *unequivocal,* and that most of the observed increase in globally averaged temperatures since the mid-20th century is *very likely* due to the observed increase in anthropogenic (human) greenhouse gas concentrations. Furthermore, the report concluded that since the release of the IPCC's Third Assessment Report,[4] the simulation models have markedly improved such that "an anthropogenic signal has now more clearly emerged in formal attribution studies of aspects of the climate system beyond global-scale atmospheric temperature, including changes in global ocean heat content, continental-scale temperature trends, temperature extremes, circulation and arctic sea ice extent." The awarding of the 2007 Nobel Peace Prize to Al Gore and the team of more than 2,000 scientists who participated in the IPCC would seem to lend credibility to these conclusions.

Yet, anyone paying even casual attention to media reports cannot help but notice that the issues are highly contested. Numerous articles, books, and reports claim that global warming is a hoax or a ploy. Not only are the potential impacts of climate change and the role of humans in the observed rise in temperature being questioned, but also whether Earth is indeed warmer relative to historic records. Just about the only thing that everyone seems to agree on is that the concentration of greenhouse gases in the atmosphere has sharply increased in the last century.

3. *Climate Change 2007: Fourth Assessment Report of the Intergovernmental Panel on Climate Change.* 4 vols. Intergovernmental Panel on Climate Change, Geneva, Switzerland, 2007.

4. *Climate Change 2001: Third Assessment Report of the Intergovernmental Panel on Climate Change.* 4 vols. Intergovernmental Panel on Climate Change, Geneva, Switzerland, 2001.

Again, the question of whom to believe comes up. We are not climate scientists, but we are fairly well informed and have enough grasp of the complex science behind it that we can see through the distortions, exaggerations, and obfuscations in many reports.

Since the question of global warming is so intimately connected with the use of energy, we will take some time to go through the three questions we posed at the start of this section. The reason to go through these arguments is that, in the end, we all have to collectively make certain value judgments, and science does not provide that. How we choose to deal with climate change has an enormous and direct impact on the lives and lifestyles of billions of people. For many it could mean the difference between one and two square meals a day,[5] while for others it could mean massive relocation to "foreign" lands where they will have to adapt to a new lifestyle and a new way of earning their livelihood. These are existential issues. No wonder then, that we have well-intentioned and knowledgeable people on both sides of the issue.

The greenhouse gas effect is real—there is no denying this fact. Had it not been for the greenhouse gas effect, Earth's average temperature would have been well below freezing at about 0°F (−18°C), instead of the very hospitable—for us humans, at least—59°F (15°C). There is also no doubt that CO_2 is a greenhouse gas and that its concentration has increased substantially over the last 150 years. Its concentration in the atmosphere prior to the industrial revolution was 260 parts per million (ppm), and today it is around 380 ppm and climbing. We are currently pumping about 24 billion tons of CO_2 into the atmosphere each year through the use of fossil fuels. If condensed into a liquid, these 24 billion tons of CO_2 would occupy about 5 cubic miles. The amount of emissions have been and continue to be on the rise. No one disputes these facts.

Arguments arise when scientists start attributing the rise in CO_2 levels to anthropogenic emissions. The consensus view expressed in the IPCC's Fourth Assessment report is that the observed rise is *very likely* due to human activity. Atmospheric scientists who question this conclusion emphasize that in order to match the quantitative rise in CO_2 concentrations with the amounts of emissions, the models have to use a lifetime of more than a century for the turnover of CO_2. In other words the CO_2 injected in the air takes about a century before it mixes with the rest of the biological and geological sinks. Many scientists contend that the lifetime for CO_2 is on the order of a decade or two.[6] In his review of two recent books on the subject, Freeman Dyson summarizes the argument for the life-cycle of CO_2 by examining the annual variation seen in the published CO_2 data. While there has been a steady increase in the atmospheric CO_2 concentration, there is an approximately 14% variation between

5. Robert B. Zoellick, Chairman. *Rising Food Prices: Policy Options and the World Bank Response*, World Bank, Washington, D.C., 2008.

6. Freeman Dyson in *New York Review of Books*, June 12, 2008, http://www.nybooks.com/articles/21494. Accessed June 2009.

spring and fall values. This seasonal difference arises from CO_2 being fixed by plants as they begin their growth in spring and then its release by the decay of plants beginning in fall. These data suggest that CO_2 would completely mix with the biosphere in about a decade. If this reasoning is correct, the emissions could not account for the observed rise, which leads one to conclude that there are other natural sources of CO_2 that the climatologists have not accounted for in their models.

Scientists also argue about whether Earth is warming up and, if so, compared to when. The IPCC's Third Assessment Report featured a graph that vividly showed a marked rise in global temperatures in the last 50 years coinciding with the equally sizable growth in atmospheric CO_2 concentrations. However, it was found that the analysis that produced the graph was flawed, and the IPCC did not use it in its Fourth Assessment Report. Others have argued that Earth has been warming at a rate of about 1°F a century since it emerged from the last ice age, and that there has been no noticeable increase in the rate of warming since the onset of industrialization. Many glaciers around the world are indeed melting, but they have been doing so for a long time, which means that there are competing explanations for almost all of the observations relevant to this discussion.

Arguments get even more complicated when we start examining the impact of the temperature rise on sea levels and the climate in general. Because reliable data for temperature, sea level, rainfall, and gas concentrations of Earth are available only for the recent period, of necessity scientists have to rely on proxy data and statistics. We will not resolve this debate here. Our point is simply to alert you to the fact that there is a scientific debate on the subject of climate change. The science is not settled. As scientists, we found IPCC's emphasis on *consensus,* which was then interpreted in the media as unanimity barring a few fringe dissenters, somewhat strange. Science has long given up on majority as the arbiter of truth.

Uncertainty surrounding the question of climate change caused by human activity aside, we still face rising CO_2 levels. Should we not engage in activities to mitigate the risk posed by rising greenhouse gases? What about taking steps to be able to adapt to the possible climate changes? Anthropogenic or not, we have to assume that there is a possibility that the outcomes could range from minimal change to as grave as described by the IPCC, or perhaps even worse. It thus becomes a question of risk mitigation, and akin to taking out an insurance policy against a potential catastrophe.

The Stern Review, a 2006 report by the British government on the economics of climate change, frames the question of climate change precisely in terms of an insurance policy.[7] It concludes that the economic risk of doing nothing

---

7. *The Stern Review on the Economics of Climate Change,* October 2006. http://www. hm-treasury.gov.uk/independent_reviews/stern_review_economics_climate_change/ sternreview_index.cfm. Accessed November 2006.

to avert greenhouse gas emissions is far greater than the cost of implementing strategies to curb them. The risk was assessed at between 5% and 20% of gross world product (GWP) per year, while the cost was only about 1% of GWP. Even with a later revision of this number to 2% of GWP, it would seem like a no-brainer to embark on a serious effort to quickly curb CO_2 emissions. However, there are also eminent economists who point out that the analysis by Stern is based on a questionable methodology of not discounting, or very weakly discounting, the future costs. Stern used this approach because of ethical considerations and the responsibility of the present generation towards future generations. However, by calculating the net present value of the future costs using market discount rates, William Nordhaus, Sterling Professor of Economics and a Fellow of the American Academy of Arts and Sciences, comes to a very different conclusion. He advocates a cautionary approach to curbing CO_2 emissions because the immediate impact on the economic welfare of society from measures to curb CO_2 emissions in any significant amount could be a lot more dire than the discounted future costs.

In taking out most insurance policies, we protect ourselves against certain risks while accepting others. That should be our approach here too. We must compare the investments needed to reduce CO_2 levels against those needed to adapt against the possible consequences of rising CO_2 levels. As with the scientific basis for anthropogenic greenhouse emissions causing global warming, we see that the economic analyses are also areas of active debate. However, instead of debate, all too often we have encountered name-calling that ends the debate: names such as *deniers* or *industry-stooges* for those opposing the idea of human-activity-induced climate change, and *believers* or *yellow scientists*—that's supposed to be a play on yellow journalism—for the proponents of that idea. We cannot resolve those controversies here, but we want to acknowledge their existence. Nor do we suggest that we debate these issues until they are completely resolved, because that would be advocating inaction.

Our stance is that even without the concerns of global climate change, meeting the future energy demand is going to pose an enormous challenge. The additional requirement for curbing greenhouse gases makes the task much more challenging. Greater use of renewable resources—or income resources as we prefer to call them—such as wind, solar, and biomass could mitigate greenhouse gas emissions, but these technologies are in their infancy and could well require decades of development. As we describe in chapter 7, while these resources have the potential to meet a significant a part of our energy demand, our current technologies for utilizing them are inadequate. Most of the alternative energy technologies are very expensive, and thus we could end up squandering resources that could otherwise be used to improve the quality of life for billions of people, including helping them reduce the impact of potential climate change by building dykes or water desalination plants. If we rush to implement current technologies at the required scale, we run the risk of creating even greater harm for some unintended consequence. "Look before you leap" is an adage that we need to follow.

The only technology that we have today to produce energy at the desired scale without greenhouse gas emissions is nuclear power. This is an option that many do not want even to consider, because there has been no clear solution to the issue of nuclear waste disposal, nor has the question of increased risk of nuclear proliferation been adequately addressed. However, we believe that leaving nuclear energy out of the equation is neither realistic nor responsible.

Inherited and Income Resources

We divide our energy resources in two categories, "inherited" and "income" resources, and use these terms throughout the book. Inherited resources, which include oil, coal, natural gas, and uranium, are finite because they were formed many millennia ago and may be exhausted within this century. In contrast, income resources, which include hydropower, biomass, photovoltaics, wind, solar thermal, tidal, and wave power, will be available to us forever barring a disaster beyond our ability to predict (or for as long as the sun shines).

Many observers, we among them, believe that (1) the world will shortly face major crises over supplies of its dominant fuel, oil, producing which at rates demanded by the economic growth is getting harder and more expensive; (2) the relatively new and emerging alternative energy technologies, including solar-generated electricity, produce only tiny amounts of energy today, are likely to take decades before they can begin to produce truly significant amounts, and are almost certain to face substantial environmental problems of their own; and (3) the world does not seem to understand, or ignores, the magnitude of the problems involved if oil production drastically slows by mid-century. We will need not only to replace much of the 1 CMO currently supplied by oil but also to find the additional energy required to satisfy the rising demand of the billions of people whose current use is far below the global average as they strive to improve their standard of living.

If we continue to increase energy consumption the way we have been over the last century, the total demand for energy by mid century (2050) would be around 9 CMO! The requirements could be lessened, if greater attention were paid both to reducing energy demand and to increasing energy efficiency. Perhaps through markedly improved efficiency and a serious conservation effort we will be able to reduced our energy demand by as much as 100% of our current rate of energy consumption. Even then, the total energy requirement in 2050 would be 6 CMO, or twice that of today. Many are placing their hopes on an unprecedented growth in the use of one or more of the income resources. The nascent energy industries such as wind and solar would have to expand several hundred to thousand times in order to deliver just 1 CMO. If we elect to eliminate the use of fossil fuels, the alternate sources will have to expand even more.

Consider the following:

- Hydropower now produces 0.15 CMO/yr, or about 1/20th of the world's energy. Its use might be increased by a factor of 2 to perhaps a factor of 4. In doing so, we would need to flood somewhere between 40,000 and 60,000 square miles of land, much of it in current or potential use, for the reservoirs to feed the hydropower plants. In addition, the use of hydropower raises both environmental and social issues.
- Replacing today's oil industry with biomass-derived energy could require 5–10% of Earth's land area, or perhaps as much as one-quarter of the world's currently available cultivatable land, to be devoted to this purpose. Agricultural and forestry activities could be expanded, but could they be grown enough to replace a significant fraction of the current oil industry without depriving the world of needed food? After all, even if the entire 80 million acres of cropland the United States currently used to produce corn were to be used for ethanol production, it would satisfy only half its annual gasoline requirement.
- Global-level wind and photovoltaic power production would require considerably less land and would not impinge so strongly on other needed activities. Energy from these sources is likely to be more expensive than today's fossil fuels. Moreover, because these income resources are not always available when energy is needed, their usefulness would be seriously limited unless large-scale, economic energy-storage technologies were also developed. Specifically:
  - If wind machines were used to generate an amount of energy equivalent to today's oil consumption, it would be necessary to manufacture and install at least a thousand 2-MW machines every week for 50 years, for a total of more than 2.5 million machines and at a capital cost in today's dollars of almost $5 trillion. These numbers for wind power are not as daunting as the challenge of finding acceptable places to locate them.
  - Alternatively, if we follow the route of installing 2-kW photovoltaic systems, like those commercially available today for individual homes, we would need nearly 5 billion such systems, each comprising ten 4 ft × 3 ft panels. At current prices, the total installed cost is likely to exceed $80 trillion. Finding the 5 billion suitably sized and located homes is another question altogether.
- Finally, the further development of nuclear power, which is already equal to hydropower as a contributor to global energy supply, is lagging. Opponents urge its abandonment, citing past difficulties encountered by the early developers and regulators of the industry, and uncertainties about the safety of storing nuclear waste for tens and even hundreds of thousands of years into the future. The equivalent nuclear power generation requirement to match today's energy contribution from oil alone would require the installation of about

2,500 of today's average-sized reactors (900-MW$_e$ rating), or in other words about 1 plant per week for 50 years, at a cumulative cost on the order of $5–10 trillion.

The task facing society as it seeks to provide for future energy needs is daunting, but continued delay in facing the facts will only make matters worse. Our choice of the words "inherited" and "income" betrays our bias in the belief that we will be better off living off our income rather than our inheritance. But weaning us from our current dependence on fossil energy is not a simple matter. Some have argued that in order to facilitate the transition, we must increase the price of fossil energy, either by taxing the carbon emissions or through a cap-and-trade mechanism. What we have to guard against is making energy affordable only for the affluent and inadvertently denying access to billions who are now aspiring to better their standard of living—those who are least able to afford an increased price for energy.

Developing nations obviously cannot be denied access to additional energy, which is key to improving their citizens' quality of life. However, we can collectively adopt lifestyles that avoid unnecessary use of energy. Such lifestyle changes to reduce energy use will be difficult to achieve, especially if they involve personal inconvenience. Decisions about installing energy-efficient equipment in industry or commercial activities will be largely dictated by economics. For example, many retail managers recognize that overall lighting in their shops could possibly be reduced, but they are also cognizant that sales are driven by well-lit displays and are therefore reluctant to reduce the lighting. Personal decisions (including purchase of an energy-efficient home, car, or light bulbs, as well as changing to a purely vegetarian diet) will depend on additional factors. It is our view that public awareness spurs motivation and is a necessary first step toward large-scale conservation or efficiency effort.

Overview of the Book

The book consists of three major parts. Part I summarizes past energy use and describes the basic elements of our current energy system, the various resources we use to meet our energy needs, and the overall pattern of their use. Currently, about 85% of our energy comes from our inherited resources. We have used historical and current data on energy as guides to project future energy use and to estimate the total energy requirements through 2050. We have chosen 2050 as the time horizon for this study because it is far enough in the future to enable major changes to occur in our energy use, while not so distant that all estimates for future consumption get too uncertain to be useful.

Two points will become evident from the discussion: (1) the energy industry is so large and is so slow to change that it takes decades to make any significant change in global patterns of production and consumption, and

(2) we have been essentially living off our inheritance, and we should quickly switch to our income resources as much as possible or face potentially drastic consequences. How we might do that and what it would take to achieve that goal are the questions that we consider in this book.

Part II examines our inherited and income energy resources. We first discuss the fossil resources of coal, oil, and natural gas, which were produced over eons and are unlikely to significantly increase with time, although new technologies may result in our ability to extract resources that are currently not economical. The discussion of our inheritance also includes the amount of uranium available for use in nuclear reactors. Because we believe that nuclear energy will play an important role in our future and because it is the most misunderstood of the energy technologies, we discuss this resource in somewhat greater detail than others. We include discussions of fuel preparation, reactor technology, and waste handling, along with a discussion of the perceptions about, and the realities of, nuclear energy. We next consider the income resources, which can last for as long as humankind is around. Under this category, we examine the potential of hydropower, geothermal, solar thermal and photovoltaics, wind, and biomass to provide for our energy needs.

As will become evident by the discussions in parts I and II, the task ahead of us is enormous and will only be exacerbated if we continue in our entrenched ways or make only small adjustments. Accomplishing anything significant generally requires a combination of aptitude and attitude. In the present case, aptitude refers to technological advances. While technological advancements and innovations are seriously needed, we note that unless we are also willing to make some attitudinal changes, the goal of living on sustainable energy is likely to elude us. In part III, we discuss the role that increased efficiency and conservation can play in mitigating the crisis. These offer the quickest means to avert the impending crisis for a period of a few decades, during which time we must gear up our income resources. In part III we also offer suggestions for planning and implementation of "new" approaches to resolving our energy dilemma. Even so, it is not until the latter half of the 21st century that we expect our income resources could become substantial contributors.

Part I

Energy Use

2

Historical Energy Development and Future Dilemmas

This chapter reviews how our energy consumption grew to its present state. This growth was often a result of a fundamental change in lifestyle that itself was engendered by a shortage of one resource or another. To summarize in a few sentences the vast field of anthropology, for a long time we humans sustained ourselves as hunters and gatherers in central Africa. Our overall population was small, and we had few worldly possessions. We kept these possessions to a minimum, for they would have only encumbered our ability to move about to locate food. Faced with a shortage of game, some of us migrated to other parts of the world, and some discovered agriculture. Agriculture brought with it a complete change in the way we lived. We settled down in one place and raised our food through an increasing expenditure of energy. We also became more susceptible to diseases caused by our own waste, for unlike in the past we were now living in close proximity with the waste and the pathogens that flourished in it. Also, material possessions became valued, which meant an even greater use of energy to acquire them. In their landmark book, *Limits to Growth*, Meadows, Randers and Meadows have traced this history and also pointed out that time and again an inexorable desire for growth, coupled with an inability to sense the approaching limits of the required resources or to comprehend the ability of sinks to absorb the waste we produce, have led society to unsustainable situations.[1]

1. D. Meadows, J. Randers, and D. Meadows. *Limits to Growth*, Chelsea Green, New York, 1972. A third edition, titled *Limits to Growth: A 30-Year Update*, was published in 2004.

Our use of fossil fuels provides examples of both kinds of limits that *Limits to Growth* discusses. On the one hand, the increasing difficulty of obtaining oil and meeting the ever-increasing demand for it is a classic case of resource limitation. On the other hand, the potential climate change caused by accumulating greenhouse gases (GHGs) in the atmosphere represents the limited ability of Earth as a sink. We have experienced limits of sustainability before, and it is instructive to learn from that history of driving forces that allowed us to transition from one fuel to another. Yes, those transitions took a lot of time, and no, they were not painless. We hope that lessons from history might show us how to minimize the pain for our transition from the use of fossil fuels.

Total Energy Use

Antiquity to the Industrial Revolution

In presenting data on historical global energy use, we have had to rely on incomplete estimates made by past investigators. In fact, data covering energy use for the entire world became available only in the twentieth century. What is clear is that as mankind emerged as a species, it existed mainly on fruits, grasses, and seeds (biomass) in quantities that provided each person with approximately 2,000 Calories. Note that one food Calorie (uppercase C) is equal to 1,000 thermal calories (lowercase c), which is a kilocalorie (kcal). Those 2,000 Calories of per capita daily food consumption is equivalent to a little less than one cup of oil per day or about 20 gallons of oil per year (GO/yr).

Over a period of thousands of years, we humans discovered the use of fire—thus using additional biomass—and started to employ domesticated animals for hunting, our annual per capita energy use increased by as much as 2.5 times (see the Total column in exhibit 2.1), to about 50 GO. About 100,000 years later, in the Middle East's prosperous Fertile Crescent, which flourished about 5,000 b.c.e., energy use per capita doubled again to around 100 GO; the increasing use of animals required more energy-intensive feeding, and the preparation of more refined foods required greater energy. About 2,000 years ago the world's population was about 250 million and was probably using energy at an average rate that was only 20% greater than the per capita use in the Fertile Crescent 5,000 years earlier. The global energy consumption at that time would have been equivalent to 30 billion GO/yr, or roughly 0.027 CMO/yr. Perhaps one-third of that energy use took the form of food for domesticated animals.

Toward the end of the Middle Ages in the 1400s, people in Northwestern Europe were devoting larger amounts of energy use to feeding animals, using considerably more energy in households, and expending measurable amounts of energy on industry and transportation. This group's overall annual energy use, an estimated 260 GO per capita, was more than twice that

Exhibit 2.1.
Estimated Energy Use through the Ages: A Brief Summary

Date years before the Present	Region	Annual per capita consumption (GO)						Population (millions)	Global Annual Energy consumption (CMO)
		Food		Home plus commerce	Industry plus agriculture	Transport	Total		
		Human	Animal						
1,000,000	Africa?	20	0	—	—	—	20	?	NA
100,000	Africa?	20	10	20	—	—	50	?	NA
7,000	Fertile Crescent	20	20	40	40	—	100	?	NA
2,000	**World**						**120**	**250**	**0.027**
600	Northwestern Europe	20	40	120	70	10	260	?	NA
130	Germany, United Kingdom, United States	20	50	330	250	140	790	?	NA
130	United States						960	130	0.035
30	United States	30	80	680	940	650	2,300	200	0.44
0	**World**[a]	**30**	**60**				**450**	**6,500**	**3.0**
	United States	35	90				2,300	300	0.63
	China						600	1300	0.22
	India	25	50				180	1,000	0.08

Historical data adapted from H. Brown, *Annual Review of Energy*, Vol. 1, Stanford University Press, Palo Alto, Calif., 1976; E. Cook, *Scientific American*, vol. 225, p. 135ff., Sept. 1971; and J. C. Fisher, *Energy in Perspective*, John Wiley and Sons, New York, 1974.
[a]Includes noncommercial energy.

of the approximately 120 GO per capita used worldwide about six centuries earlier. As we shall see in chapter 3, about half the world's population still consumes less than 260 GO annually per capita, and their living conditions are not much different from what they were centuries ago. The Industrial Revolution of the 18th and 19th centuries substantially increased energy use and, concomitantly, the standard of living, as reflected in parameters such as life expectancy, infant mortality, food intake, mobility, and gross domestic productivity, as well as simple creature comforts that many of us take for granted.

The Industrial Revolution to the Present

Up until the Industrial Revolution, per capita energy use was about the same throughout all regions of Earth and, as indicated above, chiefly consisted of energy used directly as food for humans and animals. Although some regional differences did occur because of the energy used for heating in colder regions, most energy use constituted noncommercial energy (see the box "Commercial versus Total Energy" below). Following the markedly increased use of energy due to industrialization, energy began to be sold, that is, commercialized. The noncommercial amount became a progressively smaller fraction of total energy use, and that portion has continued to diminish over the years. For that reason, in exhibit 2.1 the energy use for the last two or so centuries refers primarily to commercial energy.

The Industrial Revolution, which began in the latter part of the 18th century, involved a major shift toward mechanization that was accelerated by the availability of a convenient energy source—coal. Coal drove the steam engines that powered the new industrial machinery, and the steam engine made it easier to pump out the water from coal mines and facilitated the mining of coal. This positive feed back provided by the steam engine as well as inventions for the more effective use of coal helped accelerate industrialization. By the early 1800s, the per capita use of energy in the newly industrialized nations of Germany, the United Kingdom, and the United States had sharply diverged from that in the nonindustrialized nations (e.g., China and India). Beginning with the Industrial Revolution, global energy use grew rapidly as population increased and energy-consuming amenities proliferated. Although data are sparse, world energy use probably grew at about 1–2%/yr during the 19th century. In the 20th century, when record keeping improved, the estimated global commercial energy use grew at an average rate of 2.6%/yr, doubling about every 28 years. As shown in exhibit 2.2 energy consumption increased from less than 0.25 CMO/yr in 1900 to about 3.0 CMO/yr today—a 12-fold increase.

Coal was preferred over wood to fuel the fireboxes of the steam engines because, for a given volume, coal packed more energy than did wood. Coal was also easier than wood to move to the place where its energy was needed.

COMMERCIAL VERSUS TOTAL ENERGY

Different conventions are used in discussing primary energy use. Some analysts choose (as the authors do) to deal primarily with the traded, and thus recorded, energy of commerce. Others choose to include in their totals estimates of unrecorded (i.e., noncommercial) energy use such as wood and agricultural wastes gathered by individuals for their own use; charcoal traded in village markets but with transactions unrecorded; and electricity generated by small, run-of-river hydropower systems that are linked to individual users. Those who include unrecorded energy estimate that it amounted to almost 0.3 CMO/yr in 2000, or about one-tenth of the total estimated commercial plus noncommercial energy use. These analysts frequently project that similar amounts of unrecorded energy will be used in the future.

This book generally uses the commercial-energy-only convention, primarily because the accuracy of recorded trading far exceeds the accuracy of noncommercial estimates; moreover, the quantities of what is now considered noncommercial energy are likely to change in the future (e.g., as natural forests from which wood is taken for personal use as a source for heat are converted to wood plantations whose product is sold in recorded transactions to electric power producers or wood-to-alcohol conversion facilities). Another example is the photovoltaic energy that an individual establishment produces but that is not measured or tabulated; in such cases, however, the usual suppliers of electricity or fuel do not need to furnish as much energy to the establishment, and their decline in recorded energy consumption might be considered as due to conservation. In the same vein, some analysts (e.g., the U.S. Energy Information Agency) include in-plant heat recovery from discarded industrial waste materials as noncommercial energy, whereas others, including the authors, consider it as an increase in energy efficiency of the process involved and therefore a reduction in the prospective energy demand.

Its use eliminated the need for facilities such as gristmills, which relied on flowing water for their energy and had to be located by streams, or windmills, which had to be in a windy spot. Energy could be provided to essentially anywhere it was needed by easily transported coal. By 1870, average per capita energy use in the industrialized nations was about three times that in North-

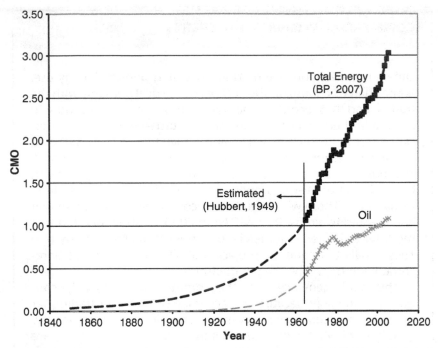

Exhibit 2.2. Global Energy Consumption since 1860 (Recent data from *BP Statistical Review of Energy, 2007*; historic data from M. King Hubbert, *Science*, vol. 109, pp. 103–109 1949)

western Europe some five centuries earlier. Per capita energy use in the United States had reached 960 GO/yr (23 bbl/yr), which was almost four times the earlier rate.

By the year 2000, the United States increased its per capita energy use to about 2,200 GO/yr (53 bbl/yr), or by another 2.3 times, where it essentially remains today. This per capita commercial energy use is about 20 times greater than that of the average world citizen of 2,000 years ago. In contrast, China had increased its annual per capita commercial energy use more than five times, while India increased its use by only 1.6 times over the estimated global average of two centuries ago.[2] Both China's and India's per capita income and energy consumption have been undergoing rapid increases since the 1990s. Between 1996 and 2003, China's energy use had grown by 8.2%/yr and India's by 5.8%/yr.

2. Overall per capita use in these two countries is larger than the commercial energy data show because both currently rely on substantial amounts of noncommercial energy.

The cumulative global commercial energy consumption since the start of the Industrial Revolution to 2006 is estimated to exceed 90 CMO for all energy forms. Of that amount, more than 30 CMO was derived from oil, for which commercial production began almost a hundred years later, in the mid-19th century. Important questions thus arise about the potential rate of growth in energy use over the 50-year period from 2000 to 2050: Will energy use increase further? Will it remain static? Could it begin to decline, and if so, when? We will address these questions in chapter 4.

To a large extent, the dilemma that we face today stems from the exponential growth in energy use. It is the property of any exponential growth that at some point it becomes unsustainable due to resource limitation. Yet, it is also the power of exponential growth that can help us out of this conundrum by allowing totally new systems to grow quickly and dominate sooner.

Changing Shares

Exhibit 2.2 shows how rapidly total commercial energy demand grew over the past century, but it does not show the major shifts in the use of energy resources or the pace of those shifts. Few appreciate how slowly these shifts in energy supply occurred and how recently oil, which now dominates global energy markets, came to its commanding position:

- The global coal share caught up with that of wood around 1880, just about the time oil began to be produced in noticeable quantities. Coal's share peaked only in the second decade of the 20th century, almost two centuries after its use became common in Britain.
- Oil's share did not equal the decreasing share of wood until about 1940, when each fuel accounted for about a one-sixth share of total energy demand, or about 0.5 CMO/yr. Oil's share did not equal that of coal until the 1960s. By that time, gas supplied about one-seventh of global energy needs.
- Hydropower, made possible in the late 1800s by the invention of the dynamo that converts mechanical energy into electrical energy, is limited in production by a lack of suitable locations. Hydropower produces about one-thirteenth of all commercial energy (thermal equivalent) today.
- Nuclear power, which was introduced commercially in the United States in 1958, had by the early 1970s grown to supply an energy share comparable to that supplied by oil in 1915 and gas in 1940. This growth was spurred in part by governments interested in the military uses of nuclear energy (e.g., France, the United Kingdom, the USSR, the United States). China, Israel, India, Pakistan, South Africa, Iran, and North Korea subsequently developed nuclear

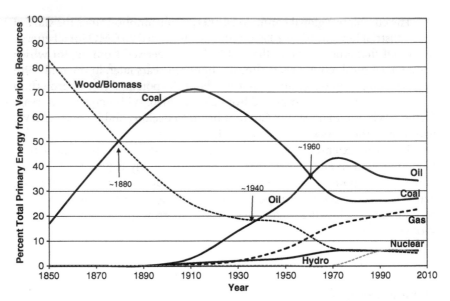

Exhibit 2.3. Changes in the Shares of Energy from Different Resources over the Last 150 Years

programs, although some were also associated with interest in weapons production. By 2000, nuclear power supplied almost as much energy as hydropower.

- The rapid gain in share by nuclear power demonstrates what can be done in a hurry when intense resolve is present. However, nuclear growth almost ceased in the 1990s and is only slowly being revived, primarily in China and India, which are experiencing rapid population and economic growth.
- Since the early 1970s, oil's share of commercial energy use has been dropping as the shares of natural gas and nuclear power have increased, with total oil production and consumption still increasing.

Exhibit 2.3 illustrates that past changes in the dominant fuel (i.e., wood to coal and then coal to oil) took many decades and suggests that similar time periods may be required for other transitions that may take place in the future (e.g., gas overtaking oil as the dominant fuel, and transitions to nonfossil resources).

Transition Times

A striking aspect of the transitions in energy share is how long they took, especially since they occurred over a period of rapid global economic expansion and population growth, which offered substantial opportunities for the use of new energy supplies. Almost three-quarters of a century was required for oil to capture just a 20% share of the commercial energy market, and almost a century for gas, in turn, to capture a similar share. Many infrastructure developments and inventions that took advantage of each new fuel's properties (e.g., the internal combustion engine and its use in the automobile, a major factor in the substantial growth in demand for oil) were required before one fuel could substitute, even in limited ways, for another. This recent history serves as a guide for analyzing the prospects for the increased use of energy-income sources, which will also require major infrastructure changes and technology improvements.

Biomass to Coal

Before the Industrial Revolution began, mankind was essentially dependent on solar energy, which provided—through crops and forests—food, fiber, fuel, and building materials. The use of coal, a more convenient energy source, contributed greatly to the transformations that the Industrial Revolution brought about. Those transformations dramatically changed the lives of people, first in Britain, then the remainder of Western Europe and North America, and finally the remaining regions. Energy use based on coal first grew rapidly in these parts of the world. Nevertheless, as shown in exhibit 2.3, it was not until about 1880 that coal finally outstripped wood as man's major energy source on a worldwide basis.

Coal to Oil

Oil was introduced to commerce in 1859,[3] but as a medical nostrum and as a lighting source to replace lighting oil obtained from sperm whales, not as an energy source. The lighting oil had helped extend the working hours of the day and increased productivity, but by the middle of 19th century the whales had been hunted to near extinction. The discovery of petroleum or rock oil happened just in time, as if on cue!

Oil was much easier to use than coal, and oil offered certain strategic advantages, as well. In the mid-19th century Britain's Royal Navy dominated the world's seas, and its ships were all powered by coal mined locally in Great Britain. However, as early as 1882 Lord John Fisher sought to switch the ships' fuel to oil. Pound for pound, oil has a higher energy content than does coal, and it afforded the ships a longer range and reduced the frequency with which they had to return to port for refueling. Furthermore, because oil could be pumped into the boiler, it would free up men from the arduous task of shoveling coal (from the bunker bays to the furnace), and they could be deployed as gunners. Lord Fisher's proposals met with stiff opposition, including from Sir Winston Churchill, on the grounds that such a switch would put the British navy at the mercy of foreign nations (mainly the United States at that time) from whom Britain would have to import oil. It wasn't until after the rise of Germany's navy that Churchill realized the value and ordered the switch. The subsequent naval

---

3. Oil was a commercial energy source in Russia about 10 years earlier, but the American discovery is generally viewed as the initiation of the modern oil industry.

victories against the German fleet during the First World War vindicated his decision and established the position of oil as the desired fuel in marine transportation.

While military advantages helped advance the technology for production and use of oil, its large-scale commercialization awaited the development of the internal combustion engine and automobiles. Initially there was a question as to whether it would be better to power cars by batteries. The higher energy density of oil won that argument over the batteries of that day and firmly established the incumbency of oil for automobiles that any alternate technology will have to overcome, unless oil use becomes a disadvantaged technology because of high price or emissions concerns.

Despite its advantages and the historic fillip it received from the military and the auto industry, oil did not surpass coal as the world's major commercial energy source until the 1960s. By that time, coal and oil supplied more than just energy; they were also being used to produce chemicals and materials. By the late 1800s coal had become the major source of dyes, drugs, and many chemicals; later oil became the source of synthetic fibers and feedstock for the chemical industry.

Natural Gas

The use of natural gas began even later, around 1880, long after coal had attained a dominant position in the energy marketplace. Synthetic gases from coal, known generally as town gas, were used earlier for street lighting in many cities. Natural gas also provides important raw materials for the chemical industry and is a source of most of the ammonia used as fertilizer. As we shall see later, the total natural gas resource is very large, and because it is a cleaner fuel than coal, considerable effort is being expended to greatly expand its use. In 2001, when the price of natural gas had fallen to around $3.00/MBtu, some predicted that natural gas use would overtake coal use. Since then, however, coal use has gained relative to gas. The primary cause is economic growth in China, where coal is the primary energy source. Natural gas continues to be "disadvantaged" vis-à-vis coal under the prevailing economic conditions.

Natural gas is a cleaner energy source than either coal or oil, having fewer contaminants. Because it has a higher hydrogen to carbon ratio, its combustion also produces less carbon dioxide (CO_2) than do the other fossil fuels—a significant benefit when considering GHG emissions and climate change. As far as GHG emissions are concerned, coal is the worst offender, and it is also the dirtiest with emissions of such contaminants as sulfur, mercury, and many other toxic elements. Petroleum use releases fewer GHGs and considerably fewer contaminants. Natural gas, with the highest hydrogen to carbon ratio, produces the least amount of greenhouse gases and is often contaminant-free (although sulfur-containing odorants are added to serve as leak detectors).

When a fuel for energy must be transported before end use, it becomes more advantaged if it takes up less space per quantity of energy content. A major challenge for increased use of natural gas is that it is often found in locations far removed from places where it can be used. As a gas, it provides

> The CO_2 emissions for the three fossil fuels have the following approximate ratios:
>
> 1.0:1.2:1.8 gas:oil:coal
>
> Thus, coal produces 80% more CO_2 than does natural gas for the same level of heat energy production.

much less energy in a given volume than do coal or oil. Coal, which is denser than oil, requires about two-thirds the volume of oil to provide equal energy. A CMO of hard coal on average will occupy a volume of $0.65\,\text{mi}^3$.[4] To deliver the energy equivalent of a cubic mile of oil (1 CMO), the volume of natural gas required at atmospheric pressure is more than $1,000\,\text{mi}^3$. By compressing the gas at higher pressures, the volume needed for the same amount of energy would be reduced. At household service line pressure, one-third as much storage space is needed. At transmission line pressure, perhaps 2 or $3\,\text{mi}^3$ is required, and as liquefied gas only $1.7\,\text{mi}^3$ is needed. However, tankage for liquefied gas is bulky and requires cooling to avoid losses, reducing its volumetric advantage.

Nuclear

In the rise of nuclear power we see a pattern of growth similar to that for oil. At first it was the military that could take advantage of its high energy content and developed much of the technology for the purpose of making a bomb. Later, it developed the technology to produce electricity for powering submarines, aircraft carriers, and icebreakers. Commercial use of that technology was made possible only because those critical technologies were in hand when demand for electricity soared in the wake of the Second World War: for air conditioners, TVs, washing machines, all-electric kitchens, and other conveniences. Further facilitating widespread adoption of nuclear power was the fact that a grid for distribution of its product, electricity, was already in place. Contrast this situation with the much talked about hydrogen economy for which we have neither a network for the distribution of hydrogen to customers nor affordable fuel cell cars that could use it. Thus, even if it were desirable to switch to hydrogen—and that is debatable—its inability to make use of existing infrastructure will slow down development of a hydrogen economy.

Each transition occurred because the newer fuel provided practical advantages and benefits that consumers valued. Coal has a greater energy

---

4. Lump coal is about 1.5 times denser than oil, but when pulverized its packing density is reduced to about 0.8 times that of oil. Thus, a CMO of pulverized coal would occupy 1.25 cubic miles.

density than wood; petroleum is easier to transport and store than is coal and also easier to use in boilers. The changes in the shares of the basic fuels were clearly not instantaneous; they required development of practical and efficient methods for each fuel's production, distribution, and use. In the case of coal, it meant development of technologies for mining, railroads, steam engines, and coal-fired furnaces. The expansion in oil use required not only technologies for prospecting and drilling of oil, but also the advent of oil-using engines, namely, the automobile. Each of these developments required substantial work and would not have taken place without the benefits afforded by the successive fuels. The historically slow adoption of new energy and other technologies provides important insights about the time that may be required for similar transitions from fossil fuels to solar and/or nuclear energy in the future. In fact, a shift from today's conventional to what are now nonconventional fossil fuels, and subsequently a shift from nonconventional fossil fuels to solar technologies represent transitions from easier-to-use fuels to what are often harder-to-use fuels and technologies. Currently, the energy these alternative sources afford is often more expensive and cannot always be produced when needed. Thus, these alternative technologies are unlikely to be adopted unless they offer compelling benefits that people around the globe begin to appreciate and value. As always, the costs of production and operation will play a significant role in their growth.

These alternative sources offer general environmental cleanliness and especially the avoidance of CO_2 emission and its impact on climate change, yet their adoption is likely to take a long time. It is difficult to predict the potential rate of change, given that in some past cases economic advantage and technical advances have resulted in rapid changes in processes. An example is the steel industry, which has undergone several "revolutionary" changes. In the first of these, almost 40 years were required for the Bessemer process, the replacement for labor-intensive "puddling," which produced poor-quality steel, to attain an equal market share. Later, it took only 10 years for the use of oxygen in steelmaking to reach a share equal to that for the conventional open-hearth air method. Some of the shortest technical transitions have involved the replacement of original materials that had adverse effects on human health with more benign materials (e.g., the replacement of lead oxide with titanium oxide pigment, and of inorganic pesticides with organic ones).

Before we can usefully discuss the requirements for changing energy sources, we must first take a closer look at the state of the energy industry today. Chapter 3 discusses the current consumption of different resources at the global level and addresses the pattern of energy production and consumption by the different regions around the world. That discussion prepares us to look at projections of what our future needs may be under several different scenarios.

3

Energy Today

The energy industry is one of the largest of the world's industries and one that directly influences the lives of the vast majority of the world's population. However, the industry's day-to-day conduct generally receives minimal public attention. Such exceptional events as an embargo on fuel shipments, a sudden rise in fuel prices, a widespread electricity shortage or outage, the rare nuclear accident, or a massive hurricane that affects oil production do make the national news, of course, and often receive prolonged coverage. Yet the more common events such as refinery fires, oil tanker wrecks, pipeline leaks and explosions, and coal-mine disasters attract the attention of only a relatively few, and then too often only in passing. And while the public attention to its activities can be fleeting, the industry is massive. Its size and influence are often overlooked, and the investments required to produce our needed energy are difficult to calculate. Using Exxon-Mobil, the largest of the petroleum companies, as a model, we estimate that the depreciated capital costs for the production of oil, gas, and chemical products derived from them are about $2.5 trillion per CMO. New investments required could be twice as large. A lack of public knowledge and the consequent lack of political will can only exacerbate our general inability to understand the enormity of rapidly changing the resources and technologies this industry employs.

We begin our analysis of the state of the energy industry by first distinguishing between primary and secondary sources of energy. Next we examine the overall production of energy by the different primary sources. We then discuss the production and consumption of energy in different regions across the globe. We also look at the per capita consumption in these regions because it is germane to the discussion in chapter 4 of the projections for future energy use. Finally, because more than 40% of primary energy is converted into secondary sources or *energy carriers* (mainly electricity) before its end use, we survey the different secondary energy sources and their markets.

Primary and Secondary Energy Sources

Primary energy consists of sources derived directly from Earth and our sun. These include sources such as coal, oil, natural gas, and uranium, as well as biomass, hydropower, wind, and geothermal power. In contrast, the secondary sources are not naturally present in any significant quantity, and they must be produced by using the energy from the primary sources. Examples of secondary sources include electricity and derived fuels such as methanol, ethanol, hydrogen, and synthesis gas.

Primary energy sources fall into two major categories, often referred to as "nonrenewables" and "renewables." As noted in chapter 1, we refer to these categories as "inherited" and "income" sources, respectively, to emphasize that inherited sources are finite and will very likely be exhausted within a few centuries, at most, unless we switch to income sources. Most of the income sources are being continuously generated, based on energy radiated from the sun or the gravitational pull of the moon, and will be available to us for many millennia.

Our inherited energy sources (discussed at greater length in chapters 5 and 6) are as follows:

1. *Fossil fuels in the form of oil, natural gas, and coal.* They are found both in very large contiguous deposits and in smaller and more widely dispersed deposits worldwide. Chemically speaking, these sources are made up mostly of carbon and hydrogen (hydrocarbons). When burned, they combine with oxygen to produce water and carbon dioxide (CO_2) and in the process give off energy in the form of heat that can be used directly. This heat can also be converted into mechanical energy in engines, which are used to drive rotary mechanisms such as car and truck transmissions, or generators that produce electricity (a secondary energy source). This simple scheme has many variations, but in essence those variations always rely on the fact that the chemical energy of the hydrocarbon fuel is greater than the chemical energy of the products of combustion—CO_2 and water.

2. *Nuclear fuels.* These fuels come from variable and widely dispersed mineral resources. Nuclear energy from these resources is released by two processes:

a. *Fission* involves splitting (or "fissioning") of certain isotopes of the heavy elements uranium and plutonium with the simultaneous release of enormous amounts of heat. Fission is currently used only to produce electrical power. Fission energy derives from the loss of mass that occurs when isotopes of uranium and plutonium are split. The energy released, which is described by Einstein's celebrated equation, $E = mc^2$, is mainly captured in the form of heat (with subsequent conversion to electricity). The same fundamental process (a net loss of mass) also occurs when fossil fuels are burned, albeit the loss of mass associated with these chemical conversions is much smaller.

b. *Fusion* involves the melding of the nuclei of low-atomic-number elements in a process that appears to be conceptually simple; however, in practice achieving controlled fusion to produce electricity has proved to be very difficult. Like nuclear fission, fusion also releases energy as the net result of a loss of mass. The fuel for fusion, deuterium—an isotope of hydrogen—is ubiquitous, and this source is potentially so large that it could last us for as long as humankind exists. In that respect, this energy source is more like an income source. Fusion reactions involving elements heavier than hydrogen could also produce energy, but such reactions are likely to be even more difficult to harness than deuterium fusion. In any event, fusion energy is unlikely to be commercially available before 2050 to 2075, at the earliest, which is beyond the time horizon of this book. However, because of its great potential, it may ultimately prove to be our largest energy resource. Therefore, developing a process for using fusion energy should be a priority objective of any energy resource development program.

Our income resources (discussed at greater length in chapter 7) include the following:

1. *Solar energy.* The various forms of solar energy available today include

a. *Hydropower,* which is derived from water released from man-made dams as well as from run-of-river electricity generators. We treat hydropower as solar energy, since it derives from solar income that drives the water evaporation and precipitation cycle.

b. *Wind,* captured by highly engineered, aerodynamic versions of windmills that convert the wind-driven motion of vanes to electricity. The potential for wind energy is substantial, and the technology has advanced significantly in recent years to be competitive with the more expensive other sources of electric power such as that produced from natural gas.

c. *Photovoltaics,* a technology that relies on semiconductors to convert sunlight directly to electricity. In a photovoltaic assembly, energy from sunlight releases a bound electron in the semiconductor (usually silicon, although other materials are also being investigated and developed) so that it can flow in an external electrical circuit.

d. *Solar thermal,* which includes the direct use of sunlight and heat, for example, in traditional brick making and for heating water or buildings, and also the conversion of solar heat energy to electricity. In practice, the latter is often achieved by concentrating sunlight onto a small area to heat a liquid, such as water, to a gas that can be used to drive a turbogenerator. In another embodiment, called ocean thermal electric conversion (OTEC), a heat engine is used to exploit the solar-driven temperature differences between heated ocean surface waters near the equator and deep waters in that region that result from the melting of ice and snow in the Arctic and Antarctic.

e. *Biomass,* which uses solar energy to convert CO_2 and water to vegetative forms that can be burned or converted to other more usable energy products. Like the fossil fuels, biomass is made up of compounds containing mostly carbon and hydrogen, but it also contains substantial quantities of oxygen. The chemical energy density of biomass is somewhat lower than that of the hydrocarbons in fossil fuels but is still considerably higher than the energy of the products of combustion—water and CO_2. Therefore, the energy released on burning a biomass source is somewhat less than that from burning an equal weight of coal. Examples of biomass use are wood burned for heat or in electric power generation, corn grain or sugar cane used to produce alcohol that can be used either as an adjunct to or as a substitute for gasoline, and vegetable oils—used as such or after conversion to biodiesel—in diesel engines.

2. *Geothermal energy.* This energy source derives from directly tapping two heat sources in Earth: (1) the heat released by Earth's slowly cooling and solidifying molten core that is at a temperature around 3,000°F (1,650°C), and (2) the heat generated by the decay of naturally occurring radioactive materials dispersed throughout our planet. The elements thorium and uranium and their decay products, as well as the mass-40 isotope of potassium (^{40}K), are the major heat sources for this component of geothermal energy. The heat that each of these materials releases will fall to one-half of its present value in 1–10 billion years, depending on the isotope. Both Earth's hot core and the decay of dispersed radioactive elements have such long lifetimes that, like the sun, they can for all practical purposes also be considered as "income" resources. Geothermal heat can be used directly or can be employed to make electricity, but both its current and likely annual contribution to the global energy mix in the near future is unlikely to exceed a few tenths of 1 CMO.

3. *Tidal power.* Tidal power results from the moon's gravitational pull on oceans. The output of the few tidal installations in use today is insignificant on a global scale. Moreover, the potential adverse environmental impacts of, and limited locations for, large-scale tidal energy use make significant contributions from this energy source unlikely, even though it might find some niche applications.

4. *Wave power.* These sources are in the very early stages of development. Systems of various kinds have been designed to capture energy from the regular up-and-down motion of the waves to drive an electric generator. One system under investigation uses wave energy to stretch electro-active polymers, which

produce an electric current as a result of the mechanical stress induced by the stretching. At present, the usefulness of wave power is limited to powering marine buoys. Future development could lead to larger systems for energy production.

Work performed by humans and certain domesticated animals can also be thought of as sources of renewable energy derived from biomass. Although the direct human energy employed to obtain food, shelter, transportation, and amenities can only be approximated, it is certainly less than 1% of the sum total of all other energy sources. On the other hand, considerable energy is consumed to produce food and animal feed. In a 2006 study, Eshel and Martin pointed out that in the United States the energy spent in producing the daily diet for an average person is on the same order as that people use for daily personal transportation.[1] As mentioned in chapter 2, a daily diet of 2000 Calories[2] is equivalent to only 20 GO per year, but there are gross inefficiencies in food production that vary widely. For some foods, such as grains, the energy needed to produce them is about the same as that which the food provides, but for others, notably red meats, the energy input may be 10 to 20 times more than what the food provides. If we look at the impact in terms of greenhouse gas emissions, the discrepancies are even larger. The fertilizers most commonly used in agriculture—nitrogen (ammonia), potash, and phosphate—are produced using energy-intensive processes, and we use large quantities of them. According to the analysis by Eshel and Martin, the greenhouse gas savings realized by switching from a "red meat" diet to a "lacto-ovo vegetarian" diet of the same total Calorie intake are equivalent to about 350 gallons of gasoline per person per year.[3] This amount is the same as what an individual driving 10,000 miles could save annually by switching from a gas-guzzling sports utility vehicle to a gas-sipping hybrid vehicle. As we shall see in the later chapters, there are few opportunities that can have an impact on greenhouse gases of this magnitude. To put the *savings* of 350 gallons in some perspective, we note that the global average per capita *total* energy consumption amounts to about 450 gallons of oil equivalent (GO). If the entire world were to adopt the "red meat" diet, total global energy consumption would increase severalfold.

1. G. Eshel and P. A. Martin. "Diet, energy, and global warming," *Earth Interactions*, vol. 10, 2006, at pp. 1–17.

2. Note that energy content of food is expressed in Calories; 1 Calorie equals 1,000 thermal calories or 1 kcal.

3. The "normal" U.S. diet includes 72% of Calories from plant-based sources. In the "red meat" diet, the remaining 28% of the Calories is provided by red meat (15%) and a mix of fish, poultry, and dairy (13%). In the lacto-ovo diet, the remaining 28% is provided by a mix of dairy (23.8%) and eggs (4.2%).

TOTAL SOLAR FLUX

The total solar radiation that reaches the outer edges of Earth's atmosphere is about 35,000 CMO/yr, of which some 12,000 CMO is reflected back into space. Of the remaining 23,000 CMO/yr, about 20 CMO is captured by Earth's global biomass systems, which include agricultural and forestry products, as well as much unused and/or unusable material such as sparsely distributed weeds and small grasses. Of this 20 CMO/yr, about 0.15 CMO is used in commercial wood energy applications, and somewhat more is found in the animal and vegetable foods we produce. Apart from biomass, we derive 0.16 CMO, which represents about 7% of total commercial energy, from hydropower. Currently, negligible amounts are also used as direct heat, solar power, and wind power. The portion of solar flux that goes unused in these activities includes that which evaporates water (which appears later as rain), unused wind, and solar heat falling on vast areas where it cannot be converted to useful forms.

Process	CMO/yr Available	Used
Evaporation-water system (hydropower)	7,800	0.16
Air-movement system (wind)	1,700	<0.02
Direct energy	13,000	
of which photovoltaic is		>0.2
Photosynthesis	1,000	
of which wood and nonagricultural production is	20	0.15[a]
Total	≈23,000	0.31

[a]Commercial energy only; noncommercial energy production may be more or less than this amount.

Our direct use of solar energy for the production of commercial fuels, such as ethanol and biodiesel, is about 0.01 CMO. In 2006, the direct use of solar energy to produce electricity with photovoltaic systems represented a miniscule 0.0005 CMO. The wind systems produce more energy than do photovoltaics, but they still provide only a very small fraction of our current energy mix. Both photovoltaic and wind energy systems are expected to provide substantially more energy in the future than they do today.

Why so very little of this incoming solar flux is captured and used is a subject of major importance to the study of global energy. To a large extent, it is due to the dilute and distributed state in which sunlight falls, and its variations with the hours, seasons, and weather. However, the fact is that we potentially have abundant solar energy. The daunting technical challenge is to find efficient and inexpensive ways to use it. We discuss this, along with other income resources, in greater detail in chapter 7.

Global Consumption of Primary Energy Sources

The total world energy consumption in 2006 (the year on which the analysis and assessment presented here is mainly based) was 3.01 CMO (as compared with 2.40 CMO in 2000, which corresponds to an annual growth of 3.8%). Oil accounted for 1.06 CMO, or about 36% of the 3.01 CMO/yr of commercial energy consumed globally. Coal supplied about 27% of the energy, and gas about 21% (0.8 and 0.6 CMO, respectively). The three other major energy sources—hydropower, wood biomass, and nuclear power—each supplied around 5% of our total commercial energy. Geothermal, solar thermal, photovoltaic, and wind together supplied less than 1% of our commercial energy. Currently about 89% of energy comes from inherited resources, with 84% of our energy today deriving from the three fossil sources: oil, coal, and natural gas. Exhibit 3.1 shows the proportion of energy obtained from the different primary sources. Exhibit 3.2 gives the absolute values for global energy production. The dotted line through the middle of exhibit 3.2 separates our inherited sources (top) from our income sources (bottom).

The 5% figure for nuclear and hydropower we quote is somewhat higher than in some compilations. The difference results from the way analysts compare electrical energy with the heat equivalent of oil energy. Strictly speaking, 1 kWh is equivalent to 3,414 Btu, but it generally requires 10,000 Btu of thermal energy from burning fuels to produce 1 kWh of electricity, which amounts to about 34% efficiency of the process. Energy sources such as hydropower and nuclear, as well as many of the emerging income energy sources like wind and photovoltaics, produce electricity directly. Therefore, 1 kWh of electricity from these sources would displace about 10,000 Btu of fossil fuel. To provide more uniform comparisons among the dominant energy sources, we treat power from these smaller contributors as if it were generated by thermal plants requiring 10,000 Btu/kWh. The U.S. Department of Energy also uses this Btu figure; BP (formerly British Petroleum), on the other hand, uses 12,000 Btu. Only the International Energy Agency (IEA) uses 3,414 Btu, which is the direct energy equivalent of 1 kWh.

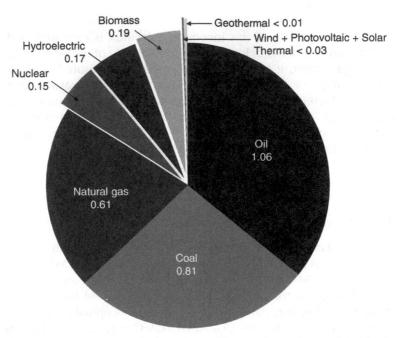

Exhibit 3.1. Proportion of Energy from Different Primary Sources (2006 data)

Exhibit 3.2.
Primary Commercial Energy Consumption in 2006 CMO/yr, rounded

Fossil Fuels			
Oil	1.06 ⎫		
Coal	0.81 ⎬ 2.48		
Natural Gas	0.61 ⎭	⎫	
Nuclear Power		⎬ 2.63 (≈89%)	
Fission	0.15 ⎭		
Fusion			*Inherited Sources*
Solar Energy			*Income Sources*
Hydropower	0.17	⎫	
Biomass	0.19	⎬	
Photovoltaics	⎫	<0.37	
Wind	⎬ <0.05		
Solar Thermal	⎭		
Geothermal	<0.01 ⎭		
Total Primary Energy	**3.0**		
(incl. nonfuel uses)			

Only the entries shown in boldface make significant contributions to global energy today.

In only a few national cases (e.g., Brazil, Canada) would the correction for nonthermal generation be significant. In compiling our data, we have converted the thermal equivalent data (which are in million tonnes of oil equivalent) to CMO and, for the tables here, adjusted them to conform to our convention of incorporating the inefficiency of thermal electricity production. With this adjustment, the energy consumption was 3.01 CMO in 2006 (excluding that from noncommercial energy sources) with a fuel input of 2.96 CMO, plus 0.05 CMO of "fuels" used for nonenergy purposes (e.g., the production of chemicals).

Uses of Primary Energy

Most compilations break down energy end-use by three main sectors: transportation, industry, and residential and commercial operations. The IEA has published estimates of the global distribution of primary energy in various end-use categories in recent years.[4] Beyond that we also need to know how much electricity was produced, and by what method, to correctly convert the raw numbers into CMO units. For those data we turned to compilations by the World Bank, which has estimated the amounts of the major forms of energy used to produce electricity.[5]

Exhibit 3.3 provides the breakdown of energy use in the United States. About 36% of the primary energy is used to produce electricity. The remaining 64% is apportioned between transportation, industry, and residential and commercial operations as follows. Transportation uses 29% of the energy, almost entirely from petroleum. The commercial and residential sector uses 13% of the primary energy, and the remaining 22% is used in industry, out of which around 5% is used as feedstock for plastics. We can take the amount of primary energy used to produce electricity and apportion it between the end uses. About 43% of the electricity is used in the industrial sector and 57% in the commercial and residential operations, with only a small amount going to the transportation sector. Adding the contributions of electrical energy to the two sectors, we note that 29% of our energy is used in transportation, 32% in industry, and 39% in the residential and commercial sector. An important point to remember from this discussion is that no single sector dominates the consumption of energy, and therefore when we look for measures to save energy, we will have to examine all the sectors and not focus on any one. The pattern for global consumption is not substantially different from that of the United States.

4. International Energy Agency. *World Energy Outlook*, Organisation for Economic Co-operation and Development, Paris, 2002.

5. World Bank, *Development Indicators*, Washington, D.C., 2002 and 2003.

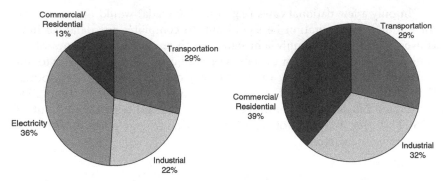

Exhibit 3.3. Proportion of Primary Energy Resources Consumed in Different Sectors of the U.S. Economy (left panel); after apportioning consumption of electricity in to other sectors (right panel)

Whereas almost all oil is used for transportation, more than 90% of coal goes for electricity production. Natural gas is used mainly for various heating tasks, for producing power, or for making hydrogen for the fertilizer and petrochemical industries. If we eventually decided to power most of our cars with electricity through the widespread use of electric cars or plug-in hybrids, we would need to increase electrical power generation by a substantial amount. Of the approximately 1 CMO of oil used in transportation, more than half is used by automobiles and light trucks. These are the vehicles most likely to be targeted for such replacement. Thus, electricity production, which currently stands at 1.2 CMO, would have to be increased to somewhere in excess of 1.5 CMO in order to replace today's cars with electric or hybrid vehicles.

Exhibit 3.4 displays the global sources of energy used to produce electricity. It gives the installed capacity in 2003 for each source as well as the amount of electricity produced in TWh and CMO units. Coal at 0.40 CMO is followed by gas, hydropower, and nuclear sources in nearly equal but smaller amounts. Oil use is significant but small compared with these other four. The other sources

Exhibit 3.4.
Global Electricity Production from Various Sources in 2003

Source/Fuel	Generation Capacity (GW)	Energy Produced		Capacity Factor (%)
		TWh	CMO	
Oil	378	823	0.05	25
Gas	1,007	2,598	0.17	29
Coal	1,119	6,160	0.40	63
Nuclear	361	2,523	0.16	80
Hydro	851	2,781	0.18	37
All Sources	3,716	14,885	0.97	46

of electricity, which primarily consist of energy income resources, contribute less than one-twentieth of the total used. The last column in exhibit 3.4 gives the capacity factor (CF) for each source, which is the percentage of the electric energy actually produced in a year compared to that which could be produced if the installed equipment operated 100% of the time (8,760 hours a year). Thus,

$$CF = \text{energy produced in a year (kWh)}/$$
$$(\text{equipment capacity (kW)} \times 8,760\,h)$$

The capacity factor tends to be high for sources that require high capital costs for installations and low operating cost of fuel. Natural gas- and oil-powered generators have low capital costs and high fuels costs and they tend to be used only at times of high demand and thus have low capacity factors.

Global Energy Business

Commensurate with the large amounts of fuel used in the energy business is the huge investment in the energy industry. The total annual cost in 2006 of fossil fuels at the point of their first entry into commerce was approximately $2 trillion; in refined, usable form the cost may have been $3 trillion or more.[6] Altogether, end-use fuels may account for nearly one-tenth of the current gross world product (GWP).

Capital investment in global energy activities is also large. Replacing the current world oil, gas, coal, and electricity production capacity could cost about $10 trillion (current dollars) or about one-fourth of annual GWP. In 2000, investment in energy supply systems was estimated to be 15–20% of total capital investment, or 3–4% of GWP. Over the past quarter century, the largest part of this investment has gone to the electric power system, of which electricity generation has been the most expensive component.

The preponderance of future energy investment is expected to follow the same pattern. Investments in end-use technologies, although less well defined, are also significant. For instance:

- The investment in energy end-use equipment may well exceed that needed for energy production facilities. Consider the investment in global motor transportation, which now involves perhaps 900 million oil-consuming automobiles, buses, motorcycles, and trucks (about 240 million in the United States alone according to U.S. Federal Highway Administration figures in 2003). Even if the average cost of the vehicles globally was only $5,000, the total amount invested in this basic

---

6. Estimates are based on 2005 U.S. costs at the coal mine mouth, oil or gas well-head, or fuel import terminal, or at the point of consumer use, as appropriate. Taxes and credits, if any, are not included in the estimates.

energy-consuming equipment of the industry would amount to more than $4.5 trillion.

- The approximately 100 million U.S. residential units alone represent an investment of as much as $550 billion in energy-consuming equipment—furnaces, air conditioning units, clothes washers and dryers, refrigerators and freezers, stovetops, and ovens.[7] In recent years, Americans spent on the order of $30 billion annually to purchase basic appliances.
- According to the 2008 edition of the *Statistical Abstract of the United States,* in 2003 the United States alone spent a total of approximately $753 billion to fuel its economy. Of this, about $178 billion went to residential uses,[8] and another $129 billion to commercial uses; $150 billion was expended for industrial purposes, and $295 billion for transportation activities. Total electricity sales were $257 billion.
- In the future, energy-related investments are likely to rise relative to those for other sectors as capital-intensive electricity demand continues to increase its share of the energy sector. In addition, newly emerging, often capital-intensive, energy-efficient end-use technologies are expected to play a larger role in energy affairs.

We need to bear in mind that these past investments have been worth many *trillions* of dollars, and no substantial change in the overall pattern of energy production or use will take place without comparable new investments. Investments on the order of $50 billion, such as those being made in nuclear power plants globally over the next decade, are substantial, but they pale compared with the total cost of the effort we face, which as we just saw runs into tens of trillions of dollars.

Regions of the World

A large World Bank database lists the energy consumption and production in 120 countries by fuel type. Because discussing each of the countries listed separately is unwieldy, most authors of books on energy, including the Intergovernmental Panel on Climate Change (IPCC), have combined the countries into groups or regions. We have chosen to do the same. Exhibit 3.5 shows our assignment of 210 countries to seven regions. Total energy production and use data are available for as many as 150 of these countries from the World Bank, BP, and the IEA, and we use their numbers in this book.[9]

---

7. Inferred from data in IEA's 2003 *World Energy and Environment Outlook.*

8. This category includes single-family and multifamily dwellings. Energy expenditures for the latter often appear in the commercial rather than the residential categories of U.S. energy statistics, and thus the costs quoted here may be understated.

9. In discussing electricity from fossil, hydroelectric, and nuclear power, we concentrate on the 65 countries listed in the 2006 *BP Statistical Review of World Energy* (http://www. bp.com/statisticalreview; accessed April 2007.

Exhibit 3.5.
World Regions, Population, and Area

Region	Population (millions)	Area (million mi²)
North America	443	9.5
Bermuda		
Canada[a]		
Greenland		
*Mexico		
St. Pierre and Miquelon		
*United States		
Central and South America	459	7.0

Antarctica	Guyana
Antigua and Barbuda	*Haiti
*Argentina	*Honduras
Aruba	*Jamaica
Bahamas	Martinique
Belize	Montserrat
Bolivia	Netherlands Antilles
*Brazil	*Nicaragua
Cayman Islands	*Panama
*Chile	*Paraguay
*Colombia	*Peru
Costa Rica	*Puerto Rico
Cuba	Saint Kitts and Nevis
Dominica	Saint Lucia
*Dominican Republic	Saint Vincent/ Grenadines
*Ecuador	Suriname
*El Salvador	*Trinidad and Tobago
French Guiana	Turks and Caicos Islands
Grenada	*Uruguay
Guadeloupe	*Venezuela
*Guatemala	Virgin Islands, British
	Virgin Islands, U.S.

Region	Population (millions)	Area (million mi²)
Europe	580	2.4

Albania	*Latvia
*Austria	*Lithuania
*Belgium	*Luxembourg
*Bosnia and Herzegovina	*Macedonia
*Bulgaria	Malta
*Croatia	*Moldova
*Czech Republic	*Netherlands

(*continued*)

Exhibit 3.5.
(*continued*)

Region		Population (millions)	Area (million mi²)
*Denmark	*Norway		
Faroe Islands	*Poland		
*Finland	*Portugal		
*France	*Romania		
*Germany	*Slovakia		
Gibraltar	*Slovenia		
*Greece	*Spain		
*Hungary	*Sweden		
Iceland	*Switzerland		
*Ireland	*Turkey		
*Italy	*United Kingdom		
	*Yugoslavia		
Russian Group		274	8.5
*Armenia	*Russian Federation		
Azerbaijan[a]	*Tajikistan		
*Belarus	*Turkmenistan		
*Georgia	*Ukraine		
*Kazakhstan	*Uzbekistan		
*Kyrgyzstan			
Asia/Pacific		3,821	11.8
*Afghanistan	Maldives		
American Samoa	*Mongolia		
Australia[a]	Nauru		
*Bangladesh	*Nepal		
Bhutan	New Caledonia		
*Brunei	*New Zealand		
*Burma	*Pakistan		
Cambodia	*Papua New Guinea		
*China	*Philippines		
Cook Islands	Samoa		
Fiji	*Singapore		
French Polynesia	Solomon Islands		
Guam	*Sri Lanka		
*India	*Taiwan		
*Indonesia	*Thailand		
*Japan	Tonga		
Kiribati	U.S. Pacific Islands		
*Korea, Democratic Republic of (North)	Vanuatu		
*Korea, Republic of (South)	*Vietnam		
*Laos	Wake Island		
Macau			
*Malaysia			

Exhibit 3.5.
(*continued*)

Region		Population (millions)	Area (million mi²)
Middle East/North Africa		325	4.3
Algeria[a]	*Libya		
*Bahrain	*Morocco		
Cyprus	*Oman		
*Egypt	*Qatar		
*Iran	*Saudi Arabia		
*Iraq	*Syria		
*Israel	*Tunisia		
*Jordan	*United Arab Emirates		
*Kuwait	Western Sahara		
*Lebanon	*Yemen		
Sub-Saharan Africa		677	8.6
*Angola	*Gabon	Reunion	
*Benin	*Gambia, The	*Rwanda	
*Botswana	*Ghana	Saint Helena	
*Burkina Faso	*Guinea	São Tome and Principe	
*Burundi	*Guinea-Bissau	*Senegal	
*Cameroon	*Kenya	Seychelles	
Cape Verde	*Lesotho	*Sierra Leone	
*Central African Republic	*Liberia	*Somalia	
*Chad	*Madagascar	*South Africa	
Comoros	*Malawi	*Sudan	
*Congo (Democratic Republic of, Brazzaville)	*Mali	*Swaziland	
*Congo (Republic of, Kinshasa)	*Mauritania	*Tanzania	
*Cote d'Ivoire	*Mauritius	*Togo	
Djibouti	*Mozambique	*Uganda	
Equatorial Guinea	*Namibia	*Zambia	
*Eritrea	*Niger	*Zimbabwe	
*Ethiopia	*Nigeria		
World Total		6,579	52

[a]Because these countries' energy use is significant, they are included in the book's energy data.

The country groupings we use differ somewhat from those of some other analysts. The principal difference arises from our perspective on current policies and alliances created by the dissolution of the former Soviet Union. For example, BP uses the category "Europe and Eurasia" that comprises Europe and the former Soviet Union, and the U.S. Energy Information Agency (EIA) lists the same countries in two categories: as belonging to the Organisation for Economic Co-operation and Development (OECD), or as non-OECD and Eurasia. Such listings were justified when the Soviet Union held sway over many nations that are now part of, or are striving to enter, the European Union. While comparisons of past and present performance may be made easier by continuation of categories created in earlier times, we have not been bound by these conventional divisions. Instead, we selected groupings that we believe more closely relate to current or even likely future trends.

In this regard, we have created a separate grouping, designated as the Russian group, that includes the Russian Federation and its closest neighbors, such as Azerbaijan, Belarus, and the Ukraine. The other members in the group previously labeled by the EIA as non-OECD Europe and Eurasia and by BP as Europe and Eurasia are those nations that are, in the main, striving to enter the European Union, and we have grouped them under Europe. Our change of grouping emphasizes the importance of the Russian group as energy suppliers, and the dependence of Europe on outside sources of energy, particularly oil and gas. The Russian group, representing a major portion of the former Soviet Union (many of its smaller components are included in our European group), is a net exporter of energy to Europe. Europe imports about a third of its oil and about half of its gas from Russia. The 10 nations of the Russian group produced more energy in 2000 than in 1980 but, conversely, consumed less. These nations are recovering from the dissolution of the Soviet Union. They are potentially unstable, with political control of some of the countries in dispute and ongoing arguments about "state" versus "private" control of energy companies. As the many economic and political problems of these countries are resolved, the group's energy consumption is likely to rise. We do not know whether the regional energy production will rise correspondingly. Therefore, the group's exports could be smaller (or larger) in future years.

We have also broken with the conventional categorizations of the Middle East and African regions by removing the Arabic nations Algeria, Morocco, Libya, and Egypt from the African category used by some others and instead joining them with their counterparts in the Middle East. These North African countries seem likely to join in strategic alliances with the Middle Eastern nations over oil and gas production policy. While on the surface most of the countries in the Middle East/North Africa region have relatively stable governments, there are potentially destabilizing conflicts involving old tribal, religious, and cultural differences, as well as rapidly growing populations that feel they are not adequately represented by the respective governments. Political disruptions could reduce exports, perhaps substantially from this region.Likewise, whereas some analysts include the low-energy-consuming nations of the Caribbean in North America, we group them with South American countries, because their pattern of energy consumption more closely resembles those of the South American countries.

Another difference is our inclusion of Australia and New Zealand with the Asian region. BP and the EIA list Australia and New Zealand under Oceania. Other analysts list Australia and New Zealand along with other OECD countries and the remaining countries of the region as Oceania. However, the energy production and consumption of Oceania is very small, and we did not feel it warranted a separate region.

The other regional reassignments, particularly those within our Asia/ Pacific region, have essentially no impact on our view of world energy inasmuch as most involve countries that do not produce or consume much energy. Similar statements can be made about the Middle East/North Africa combination. The North African nations produce very small amounts of oil and gas compared with the nations that conventionally make up the region, and their transfer from the conventional African nation assignment has generally had only a minor impact on the totals for Africa.

Finally, Greenland is a "province" of Denmark, but geographically it is part of the North American continent. Energy consumption and production in Greenland are reported as a part of Denmark and thus get included in Europe's consumption and production. Fortunately, these numbers are small and do not affect our analysis. The larger and more significant are Greenland's fossil and nuclear resources. These number are available distinct from the resources of Denmark, and we report them here as part of resources of the North America region.

Regional Production and Consumption of Energy

With these groupings in mind, in exhibit 3.6 we present the total primary energy production and consumption of each of these seven regions. The patterns of production and consumption of energy around the world are far from uniform. As mentioned earlier, energy consumption is strongly linked to the standard of living, and production is largely governed by the availability of resources. Further breakdown of the data for each of the regions by the fuel type is given in exhibit 3.7. It may come as a surprise to many, but the two largest producers of energy are the Asia/Pacific and North America regions, producing 0.71 and 0.60 CMO, respectively, in 2006. The Middle East/North Africa region is the third largest producer (0.52 CMO). However, this region stands alone in having a huge surplus of energy. It produces roughly three and one-half times the energy that it consumes. In absolute terms, it is also the region with the largest exports, which amounted to 0.38 CMO in 2006. The region with the next largest excess energy production is the Russian group; in 2006, it exported the equivalent of 0.19 CMO. Sub-Saharan Africa and Central and South America are the two other net energy producers, but the amounts of their excess energy are small—only 0.02 and 0.01 CMO, respectively. Asia/ Pacific, Europe, and North America are net energy importers. Europe and the Asia/Pacific regions each imported 0.23 CMO of energy in 2006, whereas the North America region imported 0.12 CMO in that year.

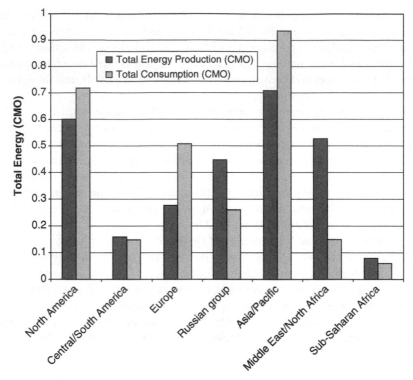

Exhibit 3.6. Breakdown of the Production and Consumption of Primary Energy by Region

Although rates of global energy production and consumption have not changed greatly in recent years, significant changes have occurred in energy sources, both in regional patterns of production and in use. Such changes are certain to continue with consumption increasing most rapidly in the Asia/Pacific region, because the ability of that region's large population to purchase energy-related goods is growing rapidly.

Looking at some of the salient features revealed by the data in exhibits 3.6 and 3.7, we note the following:

• The Asia/Pacific region produced 0.71 CMO and consumed 0.94 CMO (the most energy consumed by any region), for a net import of 0.23 CMO. About two-thirds of the energy it produced came from coal. The North America region produced 0.60 CMO, more than 25% of which was in the form of coal. It was the other large importer of energy, at a net import of 0.19 CMO for about one-quarter of the energy it used; Asia/Pacific imported one-sixth. These two regions have dominated the world energy consumption and will continue to do so for some time in the future.

Exhibit 3.7.
Regional Energy Production and Consumption (CMO) by Fuel, 2006 [data rounded]

Region	Oil		Gas		Coal		Nuclear		Hydro		Total	
	Prod.	Cons.	Prod.	Cons.	Prod.	Cons.	Prod.	Cons.	Prod.	Cons.	Prod.	Cons.
North America	0.17	**0.29**	0.18	0.18	0.16	0.16	0.05	0.05	0.04	0.04	**0.60**	0.72
Central/South America	0.09	**0.06**	0.03	0.03	<0.01	0.02	<0.01	<0.01	0.04	0.04	**0.16**	0.15
Europe	0.06	**0.20**	0.05	**0.13**	0.07	**0.09**	0.07	0.06	0.03	0.03	**0.28**	0.51
Russian Group	0.16	**0.05**	**0.20**	0.14	0.06	0.05	0.01	0.01	0.02	0.01	**0.45**	0.26
Asia/Pacific	0.07	**0.30**	0.08	0.10	0.47	0.47	0.03	0.03	**0.06**	0.04	**0.71**	0.94
Middle East/North Africa	**0.37**	**0.09**	**0.16**	0.07	<0.01	<0.01	—	—	<0.01	<0.01	**0.53**	0.15
Sub-Saharan Africa	**0.08**	**0.02**	<0.01	<0.01	<0.01	0.03	<0.01	<0.01	<0.01	<0.01	**0.08**	**0.06**
Total	1.01	1.01	0.67	0.67	0.81	0.83	0.17	0.17	0.21	0.16	2.85	2.83

We estimate that the global biomass energy production for wood based on World Energy Council figures is around 0.15 CMO. The production of biofuels such as ethanol and biodiesel is much less than 0.01 CMO. Boldface entries indicate either surplus or deficit of more than 0.02 CMO, and hence the need for significant interregional transport.

- The relatively large amounts of energy consumed in North America and Asia/Pacific were apportioned quite differently among their respective populations of 443 million and 3.8 billion: the respective per capita consumption rates in the two regions were 1,800 and 270 GO (gallons of oil), reflecting the vastly different regional economic standards.
- The third largest producer of energy was the Middle East/North Africa region. It produced 0.52 CMO, all from oil and gas because this region has essentially no coal. The region produced as much energy as the combined energy produced from oil and gas resources in Asia/Pacific and North America.
- Europe produced about one-half as much energy produced in the Middle East/North Africa region, and its consumption was slightly more than the production of the Middle East/North Africa region, for a net import of 0.23 CMO. Coincidentally, the amount imported by the Asia/ Pacific region was also 0.23 CMO.
- The Russian group produced about 85% as much energy as the Middle East/North Africa region and exported about half as much; its net exports amounted to 0.19 CMO.
- The combined production of energy in the Central/South America and sub-Saharan Africa regions was roughly one-half that of the Russian group. Because the two regions' consumption rates were still smaller, they were small net energy exporters.

Inter- and Intraregional Transfer of Energy

The data in exhibit 3.7 indicate the regions with large imbalances in the production and consumption of a given energy source. The bold numbers indicate the resource for which the imbalance is greater than 0.02 CMO and hence need for interregional transfer. Most of these imbalances are in oil and gas resources, so most interregional energy transfer is for those fuels. The electricity produced by nuclear and hydropower systems is not shipped over the long distances as would be required for interregional transfer because, as a general rule, losses during transmission render power lines longer than about 600 miles (1,000 km) uneconomical and impractical. Thus, for nuclear and hydropower sources, the amount of energy produced in any region is virtually the same as the amount of energy consumed therein. Exhibit 3.7 also indicates that little *interregional* transfer of coal occurs, mainly because this resource is found in most regions. Nonetheless, substantial *intraregional* transport of coal does take place. In this regard, exhibit 3.7 does not show the huge quantities of coal that are traded intraregionally, such as from Australia to Japan within the Asia/Pacific region.

Oil and gas account for about 56% of the world's commercial energy use, and the Middle East/North Africa countries are the principal suppliers.

The dominance of this region, which supplies energy to the Asia/Pacific, Europe, and North America regions, can create international tensions as major oil- and gas-consuming nations vie for influence and/or control over the region's present and future production. Not only does the Middle East/North Africa region hold a dominant energy position, but it is also fraught with potential uncertainties, as is the Russian group; however, the Russian group is less likely to undergo disruption in the near future than is the other region.

Energy imbalances among regions substantially influence world geopolitics. Although Europe is only the third largest energy importer in absolute terms, it imports a larger fraction of its energy than either the Asia/Pacific or North America regions. European countries are thus particularly susceptible to the vagaries of their foreign suppliers. The significant U.S. dependence on imported Middle Eastern oil also makes it vulnerable to changes in that region's politics; however, the availability of intraregional imports from Canada and Mexico makes U.S. dependence on this source somewhat less critical than is the case for Europe. This point is often lost in the discussion of U.S. oil imports. Currently, imports from Canada and Mexico alone account for 35% of all U.S. petroleum imports, whereas imports from the Middle East/North Africa region account for 25% of the total.

Strategic control of energy and fuel resources is nothing new. It was a major factor in many conflicts in the 19th and 20th centuries. Japan's attempts to secure control over Asia-Pacific resources, including energy sources, led to invasions of China in the 1930s and of Southeast Asia in the early 1940s, and its entry into World War II. Daniel Yergin's book *The Prize*[10] presents a detailed account of how most of the wars in the 20th century were won or lost by the side that managed to control oil resources.

Some energy resources are transferred intraregionally. As mentioned above, Australia transfers coal and gas to other nations in the Asia/Pacific region, notably to Japan. Japan imports about four-fifths of the energy it uses, with fuels coming from both intra- and interregional sources. In the North America region, the United States is the largest importer of energy. In addition to its oil imports from Canada and Mexico, the United States also imports substantial quantities of natural gas and some electricity from Canada.

Regional Per Capita Energy Consumption

In addition to population, a region's standard of living influences its total energy consumption. Exhibit 3.8 shows the per capita energy consumption

10. Daniel Yergin. *The Prize: The Epic Quest for Oil, Money and Power*, Free Press, New York, 1992.

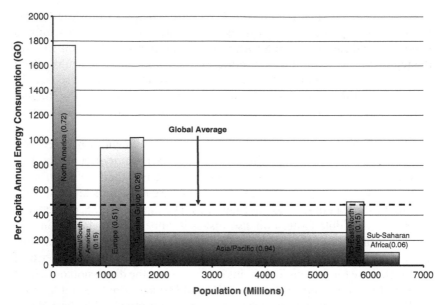

Exhibit 3.8. Per Capita Annual Energy Consumption and Population of the Different Regions in 2006

in the regions. The height of each bar gives the annual per capita energy consumption in gallons of oil equivalent (GO), and the width of the bar for each region is proportional to its population. The area of each rectangle is proportional to the total energy consumption of the region (in CMO units), and that value is also indicated within each bar. The annual per capita energy consumption in 2006 ranged from 1,800 GO in North America to slightly less than 100 GO in sub-Saharan Africa. For reference, the global average per capita energy consumption in 2006 was 450 GO.

As disparate as these regional averages were in 2006, the national differences in per capita use were greater. For example, the United States consumed about 2,200 GO/yr per capita, and Canada's per capita consumption was even higher at about 2,700 GO/yr. Mexico, on the other hand, consumed only 400 GO/yr per capita—for a North America regional average of 1,800 GO/yr. In the Middle East/North Africa region, the variations were even more dramatic. At the high end of consumption were Qatar and the United Arab Emirates, with per capita consumption of 6,800 and 3,600 GO/yr, respectively. In contrast, in Egypt and Algeria, the per capita energy consumption was only 200 and 280 GO/yr, respectively. The regional average figure of about 500 GO/yr per capita masks these national disparities. In sub-Saharan Africa, the average per capita consumption was only 100 GO of commercial energy, but in South Africa it was about eight times that figure.

This wide disparity in energy use is illustrated in exhibit 3.9, which displays per capita energy use for 64 countries for which energy data were avail-

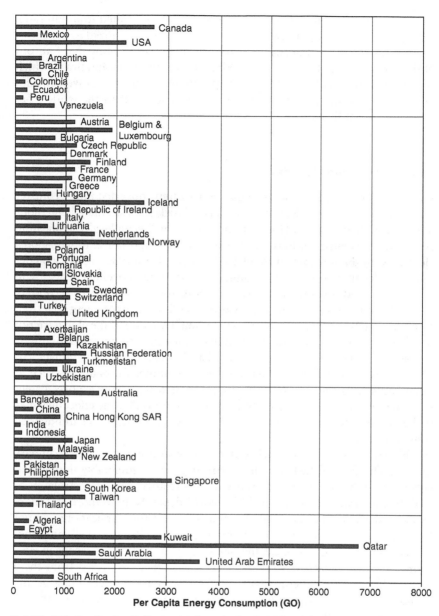

Exhibit 3.9. Per Capita Gallons of Oil Equivalent Used in Selected Countries

able from the BP compilation. The population numbers were taken from the *World Factbook* of the U.S. Central Intelligence Agency.[11] Nations with per capita consumption greater than that of the United States are either small

11. U.S. Central Intelligence Agency. *World Factbook*. http://www.cia.gov/library/publications/the-world-factbook/. Accessed October 2007.

with highly developed commercial centers (Qatar, the United Arab Emirates, Singapore, Kuwait, Belgium, and Luxembourg) or industrialized nations with frigid winter weather (Canada and Norway). It is worth noting that four out of the five most populous nations—China, India, Indonesia, and Brazil, which together account for 47% of the world's population—have a relatively smaller per capita consumption of energy, significantly lower than the world's per capita average at this time.

Secondary Energy Sources

As noted above, primary energy resources are often converted into secondary forms of energy such as electricity and derived fuels such as ethanol and biodiesel before their end use. However, the use of the "secondary" classification should not be construed to mean they are of less importance. In fact, global electricity production grew at about 2.7% per year from 1980 to 2003, a higher rate than the approximately 1.6% annual growth rate in overall energy use in the same period. Expressed in thermal terms, the current use of secondary energy is equivalent to more than two-fifths of all primary energy. Derived secondary fuels make up only an extremely small fraction of the global energy, but they are receiving increasing attention because of their potential for offsetting substantial quantities of imported oil and thus for serving national security interests. These fuels also represent opportunities for countries such as Brazil, Indonesia, and others in the tropics to increase energy exports. Lately, secondary fuels have also been getting a lot of attention for the effect they may be having on global food prices, loss of ecosystems, and other unintended consequences.

Electricity and derived fuels differ in one important way as secondary sources. The derived fuels take the form of gases or preferably liquids that generally can be stored and transported with relative ease like petroleum or coal. In contrast, electricity, which the world depends upon more and more because of its convenience and cleanliness in end-use applications, cannot be stored directly. Storing electricity requires it to be transformed into other forms of energy (e.g., chemical, mechanical, or, in the future perhaps in electromagnetic or capacitive storage systems) and then recovered from these forms when needed. Energy losses of 10–20% or more (depending on the technology used and the storage conditions) can result when transforming stored electricity into a useful form.

Electricity

As a secondary energy source, electricity is extremely useful and desirable. With a simple flick of the switch, it is made available to perform a variety of tasks: provide light, heat or cool the home, cook meals, turn motors of sizes ranging from microscopic to gigantic, and run computers and devices for communication, music, or entertainment—the list is endless. Putting emissions during *production* aside, electricity does not produce waste apart from some unused heat in *performing* these tasks at the site of the user. In that limited sense, electricity is a clean source of energy. That attribute along with its ready availability are major benefits for the customer. No wonder, then, that the growth rate of electricity use is substantially higher than that of overall energy.

Only in the late 19th century did electricity emerge from its infancy, with electricity generators and motors becoming routine items used in factory production in the early 20th century. Key early events in the history of electricity include (1) hydropower generation in the mountains of northern Italy for transmission to power streetcars in Milan in 1870; (2) the first U.S. commercial power station, supplying street lighting to San Francisco in 1879; and (3) the opening by Thomas A. Edison in 1882 of a fossil-fueled electricity-generating plant in downtown New York City to service about 400 electric light bulbs that he had installed in local buildings. (The plant efficiency was one-quarter to one-fifth that of today's coal-fired steam-generating units, and perhaps one-sixth to one-eighth of new gas-fired combined-cycle systems that use a combination of gas turbines and steam turbines.) The Chicago World's Fair of 1893 also made extensive use of electric lighting and even used electricity-driven boats to carry visitors through its canals. About the same time, California entrepreneurs began exploiting hydropower to drive gold and silver mining and refining equipment.

Today's vast electric grid began as isolated pockets that grew and merged over time, and that pattern of growth affected the way power is transmitted and, as we describe below, has ramifications for the implementation of new income or renewable sources of energy. The grid was developed with the aim of providing electricity cheaply, reliably, and ubiquitously, and in that regard it has served society reasonably well.

In the 1880s Thomas Edison and George Westinghouse debated the best method for transmitting electric power. Edison favored the direct current (DC) mode while Westinghouse advocated the alternating current (AC) mode for transmitting power. The distances over which the power needed to be transmitted were generally short at that time, around 30 miles and often less. Over these distances the AC mode of transmission suffers less loss because AC power can be transformed easily to higher voltages, and higher voltages translate to lower power losses. AC mode thus became the preferred mode and subsequently grew into the vast network that we currently have.

At high voltages AC current can leak to the ground; to minimize these losses, high-tension power lines, typically 400 kV, are strung about 100 feet above the ground. Even so, line losses make the AC mode impractical for transmitting power for distances longer than about 600 miles. Power transmission lines typically cost about $5 million per mile, including acquiring easement that could be a couple hundred feet wide. Estimated capital expenditure of about $5 billion for a 1,000-mile line would make electricity from a single power plant, whether fossil, nuclear, solar, or wind, too expensive, but sharing a line among a combination of them could make them competitive sources of power. The transmission of DC power does not suffer from these leakage losses and therefore is better suited for transmitting power over long distances. However, because the world is now wedded to AC-driven appliances, DC electrical energy needs to be converted to AC before it reaches its end users.

Electricity Production

In 2006, the world generated and used more than 19 trillion kWh of electricity, the heat equivalent of about two-fifths of all energy used in that year, or about 1.2 CMO—somewhat greater than all oil use. The major electric-power-producing countries were led by the United States, which accounted for 22.4% of global electricity production in 2006. Other major electricity producers were China at nearly 15%; Japan, 6%; the Russian Federation, 5.2%; and India, 3.8%. Together these five countries accounted for more than half of all reported electricity generation in 2006.

The role of electricity in energy affairs is likely to increase in importance over the next few decades. In 2000, only 90% of the world's urban population and less than 60% of the rural population had access to electrical systems. The developing nations in general are poorly served by long-distance electrical transmission systems. Improving the economics and external help can speed extension of these systems, and the extended use of small stand-alone mechanisms that provide electricity such as photovoltaic installations, windmills, and small hydropower units should increase the proportion of electricity use in the developing nations and areas of low population density. In the developed countries, the projected introduction of massive fleets of electric or plug-in hybrid vehicles could lead to a world where the electrical energy component increases from its current level of 40% to −50%, or even 60%, of total energy use. We therefore believe it important to discuss electricity production and use in some detail.

Electricity can be generated in different ways using various primary energy sources. By far the largest amount is produced from coal, followed, in order, by gas, hydro, nuclear, and oil. Electricity is generated when a metallic wire is moved in a magnetic field. In practice, electricity is often produced in generators that consist of rotating coils of wires surrounded by magnets. The same principle is commonly employed in exercise bicycles: the pedals turn a generator, and the level of effort is adjusted by drawing the electric current generated.

The widespread availability of electricity is remarkable, given that it must be generated at the instant it is needed. In addition, with the rapidly expanding international electronics and computer industries, increasing demands are continually being placed on electricity in terms of reliability of supply and quality, including stability of voltage, frequency, and waveform. All of these factors are influenced not only by the highly variable pattern of demand for electricity, but also by the characteristics of the generating system itself.

Some exercise machines let you see how much energy you are producing as well as how many Calories you have burned. When I (R.M., a reasonably fit "middle-aged" man) am exercising vigorously, and I mean vigorous as in huffing and sweating profusely, I produce around 125 W, and it is a rate I can sustain for only a few minutes before dropping back to a more sustainable 100 W. However, I will need to maintain that burn rate for 10 hours to produce 1 kWh of energy, which my local utility, Pacific Gas and Electric, sells to me for 12 cents.

Now, in a typical 30-min workout session, I burn around 150 Calories, which means that I would burn about 3,000 Calories in the 10 hours needed to produce 1 kWh, which is the equivalent of only 860 Calories in electrical energy. This exercise illustrates the inefficiencies of energy conversion systems; in this example I was being only 28% efficient, and I used the remaining 72% for my own existence. The exercise also points out the futility of generating any sizable amount of energy by tapping into the "human power" from fitness centers, an idea that some have espoused.

Having a large grid with several suppliers and many users provides the inertia that ameliorates some these problems caused by fluctuations in demand and supply.

It may come as a surprise, but the grid also helps reduce the overall generating capacity, because it can average out the demand over the large system. Utilities typically count on providing only 2 kW of power per household even though there are many appliances in a house that consume more power than that. Utilities rely on the fact that not all these appliances are likely to be turned on at the same time or, at least, not everyone served by the system is going to have them all on together. The law of averages comes in handy. In a large set, the probability for significant deviation from the mean is reduced. If the power were generated and distributed on a local level, the rated capacity of the system would have to be larger to account for the greater likelihood of significant departures from the average value. Proponents of more localized, distributed electricity production often describe it as the wave of the future; rarely do they mention the increased generation capacity that such a system would require.

Two relatively new income sources, photovoltaics and wind power, are pure electricity producers. A common problem they both face, however, is lack of availability *when* they are needed or *where* they are needed. This stands in contrast to the electric energy produced using inherited fuel; the energy source or fuel for these can be moved to where and when needed. Wind power is most often generated in AC form for feeding into the existing electrical delivery system. If the demand for power is absent at a time when the wind is blowing, some sort of storage device is needed or the wind turbines simply need to be turned off. Potential systems for storage that could use the AC power directly include flywheels and compressed air. Storing the energy in batteries or converting it into hydrogen through electrolysis of water would require the power to be first transformed to DC. However, those processes result in energy losses of 20% or more when carried out in the small units associated with wind farms.[12] In contrast, photovoltaic power is produced only in DC form and thus does not require transformation before being stored in batteries or used to produce hydrogen; however, it must be converted to AC to supply the more common electricity transmission and delivery systems. Theoretically, the world's entire electrical system could be converted to the initial production of DC, as Edison wished when he first commercialized electric power in New York City, but the high cost and lengthy time required for complete conversion from AC to DC make this option highly unlikely.

Demand Variation

Demand for electricity varies from season to season and during each hour of the day. In developed nations, the existing generation, transmission, and distribution systems routinely supply power reliably. These systems typically depend on the flexible use of fossil fuels and have been tailored to meet the instantaneous and rapidly changing demands of a large variety of individuals and industries. As mentioned above, developing nations with incomplete, insufficient, or overloaded transmission systems are promoting the use of small-scale electricity generation using energy income sources such as run-of-river hydropower and photovoltaics. The latter system is sometimes coupled with a storage system such as a rechargeable battery.

Exhibit 3.10 displays a typical "load-duration" curve for a U.S. electric utility. It shows the number of hours during a year that the utility provides specific levels of power to its customers. It does that by having a diverse set of power generation units, some of which are operated for most of the time, while others may be operated on an as-needed basis:

- Utilities maintain a typical excess capacity or "reserve margin" in case of needed servicing or experiencing failure of at least one and possibly two or more of their major units. The "ideal" excess capacity required can be lowered to the extent that multiple utilities are interconnected by transmission lines, enabling shared electricity production.[13] Less than 1% of the energy produced over a year taps into this capacity.

12. The efficiency of the electrolysis process can approach 100%, but only if heat is used to raise the temperature of the electrolytic medium.

13. With deregulation in the United States and other nations, the traditional practice of having a single, integrated electric utility generate, transmit, and distribute electricity to customers has changed. These three functions may now be separate and under independent control. In this case, the distributing company's reserve margin is related to the quantities of

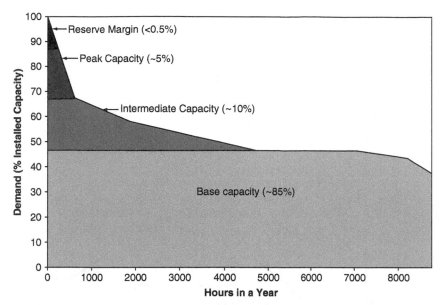

Exhibit 3.10. Typical Annual Usage of Installed Power Generation Capacity

- The segment labeled "peak capacity" is ordinarily served by small-capacity, low-capital-cost units with high fuel costs—typically oil- or gas-fired turbines. Peak load accounts for perhaps 5% or less of the electric energy delivered during the year in the United States (If the peak load occurs during daylight hours, it could also be served by photovoltaic or solar-thermal power systems.) Wind power, if available, could also fill this need and could reduce the need for traditional generation in the other categories, as well. However, wind power and photovoltaics cannot always be "fired up" at the time of need; thus, they cannot replace more than a small fraction of the traditional generation units that are installed in today's power systems without adequate provisions for large-scale storage.
- Reserve margin and peak capacity together comprise about one-third of the installed capacity, and they are unused for 90% of the time. They are often served by gas-fired power plants.
- In contrast, the base-load (relatively steady state) capacity is typically served by large-capacity, high-capital-cost units using cheaper fuels. Coal and nuclear units are primarily used to serve this component in countries that have them. Base-load electricity accounts for more than 70–80% of annual U.S. electricity production. (Utilities in other countries have different, although similar, patterns of demand.)

electricity it can purchase from generator companies and that can be delivered by the transmission companies. (Long-term contracts for generation and transmission could fill the traditional role of the "reserve margin" for the distribution company.)

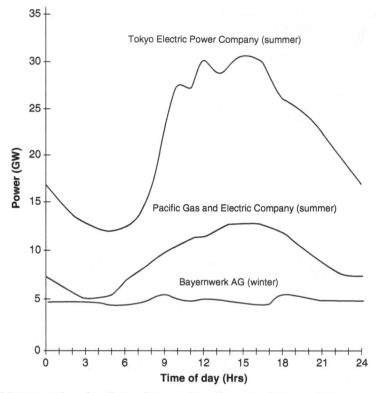

Exhibit 3.11. Selected Daily Load Demand Profiles on Peak Demand Days

- Power production units used for intermediate-load service vary widely and often include less efficient, older base-load units whose capital costs have been amortized.

The daily variation in demand for electric power places additional stress on the energy delivery system. We briefly discuss three representative examples of these daily variations from different parts of the world. Typical data are shown in exhibit 3.11 for Tokyo Electric Power Company (Japan), Pacific Gas and Electric Company (PG&E, California), and Bayernwerke AG (Germany). A major distinction among these three systems is that the first two have large but distinctly different summer peak patterns, whereas Bayernwerke's peak demand in daytime usage, which occurs in the winter, varies by only a small amount from other times of the day.

The multiple Tokyo summer peaking results from a large demand for office air conditioning and lighting. Demand is also noticeably modified by energy-conscious office workers who turn off lights and appliances when they leave their offices at lunchtime. Because the Tokyo metropolitan area is also served by a large electricity-driven commuter train system, some of the multiple peaking shown results from the morning and evening commutes. The PG&E summer afternoon peak reflects combined agricultural demands for irrigation and urban demands for air conditioning in a large, hot inland valley. The Bayernwerke winter peaking is notable because the peak demand is only slightly larger than the demand at other hours as the result of a systemwide use of ceramic resistance heaters in homes; the heaters are charged during what, in other systems, would be the off-peak nighttime hours. These heaters subsequently release the stored energy for room heating throughout much of the day.

While the Bayernwerke electricity generating system has little need for peaking capacity, the Tokyo Electric Power system draws heavily on peak- and intermediate-load generators to meet its massive swings in demand. In the Tokyo case, photovoltaic systems, which produce the largest amounts of power between perhaps 11 A.M. and 6 P.M., would be most useful. The PG&E system has peaking needs that fall between those of the other two utilities, which necessitate greater emphasis on intermediate-load power generation.

Different profiles of daily demand variation require different management approaches. Photovoltaic and wind systems are potentially well suited to meet PG&E's summer peak loads. Indeed, these electricity-generating systems are being employed in central and southern California. Photovoltaic systems benefit from the abundant sunlight, and wind systems benefit from wind movements driven by the frequent large afternoon temperature differentials between California's sun-soaked valleys and the cooler Pacific Ocean. Tokyo Electric Power Company now uses photovoltaic power to meet the summer peak demand that occurs during midday, having successfully achieved this without major changes to its overall energy management. If Tokyo Electric Power Company used solar energy sources more widely, it would need to develop large-scale electricity storage technology, discussed below, to simplify the management of the large and rapid changes in demand that characterize its system. It currently relies on nuclear power for much of its base load supply.

In general, photovoltaic and solar thermal power technologies (without electricity storage) could be especially effective as electricity producers during summer daylight hours when system demands (e.g., PG&E, Tokyo Electric) and solar inputs are also high, as in the summer peaking cases of exhibit 3.11. However, they would be much less effective if peak demand occurred during winter periods when solar inputs are relatively low.

On the other hand, although less sunny areas such as northern Europe cannot use photovoltaic-generated electricity as effectively, the wind power available there is already being used to replace some of the region's intermediate and, perhaps, base-load generating capacity.[14] In the United States, about

---

14. We note that even though photovoltaic systems are not particularly economic under prevailing conditions, the German government has promoted the use of these systems to the extent that they are making a substantial contribution to the country's electric power use.

two-thirds of the electricity produced is used to supply base-load needs. Studies conducted for the Electric Power Research Institute suggest that, under favorable solar conditions, U.S. utilities in the Southwest and West could use photovoltaic power to satisfy about one-fifth of their electricity needs without economic penalty. However, because photovoltaics would supply less power to utilities in other regions, the total contribution of this source to U.S. power might be only 5–10% at most, albeit without substantial additional costs to consumers.

Wind systems are not strongly linked to the diurnal cycle and therefore may contribute electricity to a system during many, if not all, hours of the day or night. In the windy Midwestern United States, winds are most likely to be strongest during the winter and spring, and at night—periods that do not ordinarily have peak electricity demands. Thus, wind turbines may be considered equivalent in function there to intermediate-load power generators. In Denmark and Germany, and to a lesser extent in other northern European countries with west-facing coastlines, wind has become a substantial contributor to the nation's electricity demand.

There have been many calls for doing away with coal-fired power plants altogether and replacing them with power from income sources, wind and solar. However, the intermittent nature of these sources means that they must be backed up with either gas-fired or other power sources. Our present grid system could not function on wind and solar power sources alone. In the present system, should the demand exceed the supply, the voltage would drop across the grid, and motors and machines would begin shutting down. Such a situation could be averted only with provision of a large electrical storage system. The storage device may be either a compressed air system or a new yet-to-be-invented high-tech device made of nanomaterials. Another way around the problem is to use the capacity in the grid in intelligent ways, something that could be enabled with a smart grid that has ability for a two-way communication between customers and utility providers. In the event of high demand, the smart grid would notify customers who could then choose to cut nonessential functions, such as air conditioning or heating, or face the choice between a brownout or higher charges for the service. Even more important, the grid itself could exercise that choice by reducing—instead of eliminating—the load of heating or air conditioning by changing the temperature setting, so as to avoid a brownout. When perfected, the smart grid will be able to choose for the customer the best time for running certain appliances and even reduce the customer's electric bill. A smart grid would open up a number of possibilities for saving energy, yet its real benefit lies in obviating the need for building additional generating capacity. A smarter grid could operate well with a much lower overhead capacity because of its ability to shave peak demand and shift load to periods of lower demand.

A vulnerability of the current grid is that a short circuit at one location can escalate into a major power outage affecting large areas. Such an instance occurred in August 2003, when a power line had a short circuit, and through a cascade of events led to a blackout affecting most of the northeastern United States and parts of Canada for several days. While this incident did not result from a malicious act by any individual, it points to vulnerability in our current system to sabotage. A smart grid would be able to isolate the problem areas quickly, and prevent them from spreading. On the other hand, a smart grid would also increase the risk of cyber attacks on the grid because every meter would be a potential point of access. The public is generally unaware of these issues, and the risks and benefits of a smart grid need to be more widely discussed.

Derived Fuels

Webster's *Ninth New Collegiate Dictionary* definitions of a fuel include (a) a material used to produce heat or power by burning, and (b) a material from which atomic energy can be liberated, especially in a reactor. According to this definition, wood (i.e., biomass) and fossil fuels, which can essentially be ignited with a match, obviously belong to this class, as does nuclear power, although the "burning" of nuclear fuels requires a far more exotic process than does lighting a fire. Given the substantial possibility of a decline in traditional fossil fuel resources before the end of this century—or sooner—the question arises: Which other fuels could be manufactured (i.e., derived fuels) rather than taken directly from the ground?

Hydrogen, ethanol, biodiesel, and methanol are the major derived fuels being promoted by some as future conveyors of energy from primary sources to end use. In one sense gasoline, diesel, and jet fuels are also secondary inasmuch as they too are produced by refining crude oil. However, because they are so closely related to petroleum, they are generally considered primary fuels. In a similar vein, we have included petroleum-like liquid fuels produced from coal or natural gas among primary fuels. Those fuels account for only a small portion of the fuel use globally but are significant in South Africa, where Sasol has been using the technology to produce oil from coal on a commercial scale for more than 50 years. Conversion of coal to liquids is likely to become an important technology in the future as more countries explore alternatives to petroleum-based liquid fuels.

Exhibit 3.12 shows major paths for production of methanol, ethanol, and hydrogen. Although these are clean-burning fuels and can be used effectively in many applications, some of their salient physical characteristics, shown in exhibit 3.13, may seriously limit the number of energy-using applications they can easily service. For instance, when energy per unit volume is a major factor (e.g., in mobile applications), the secondary fuels, particularly hydrogen, are at a major disadvantage when compared with oil. As the boldface entries in exhibit 3.13 show, oil has by far the largest energy content per unit volume, and hydrogen (at normal atmospheric pressure and temperature) the lowest. A gallon of petroleum contains more hydrogen than a gallon of liquid hydrogen does!

Hydrogen

Hydrogen has received much attention lately as the fuel of the future. It can be burned in a traditional internal combustion engine, albeit with some engine modifications, for motive power; or it can be used in a fuel cell to generate

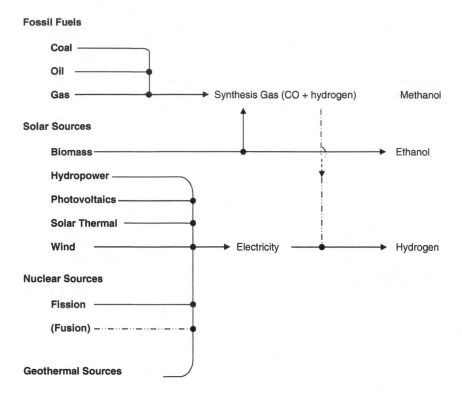

Exhibit 3.12. Paths to Major Secondary Fuels

electricity, which can then be used for transportation or other applications. The fact that the only product produced when hydrogen is burned is water has great appeal. Furthermore, hydrogen's potential use in fuel cells, which generally have higher efficiencies than internal combustion engines, also strengthens the case for hydrogen's use as a clean efficient fuel.

A fuel cell is an electrochemical device that converts the energy in hydrogen (or some other fuels such as natural gas) directly into electricity, with conversion efficiencies ranging from around 40% to more than 60%. Compare that with the typical thermal power plant whose conversion efficiencies are generally less than 40%.[15]

15. New, large, gas-fired, combined-cycle electrical generating systems based on fossil fuels have efficiencies approaching 60% (or about 80% of theoretical thermal efficiency) through the use of advanced materials in turbine blades and sophisticated heat recovery systems. Small fuel cells burning hydrogen have the potential to approach or slightly exceed this level of performance. (If the entire quantity of waste heat the cell produces could be used economically, its efficiency could also approach 80%.)

	Material	Carbon/ Hydrogen Ratio	Form	Energy Content (Btu/gal)[a]	(Btu/lb)
World's original energy source	Wood	≈0.5	Solid	29,000	8,400
Fossil Fuels (Significant energy sources since the early 1800s)	Coal	≈0.8	Solid, hard	77,000	16,500
			Solid, brown	33,000	9,000
	Crude oil	≈0.5	Liquid	140,000	19,000
	Natural gas	0.25	Gas, atmospheric pressure	130	20,300
			Gas, 6,000 psi	52,000	
			Liquefied (−95°F [−70°C])	58,000	
Secondary Fuels	Ethanol	0.33	Liquid	76,000	11,600
	Methanol	0.25	Liquid	57,000	8,600
	Hydrogen	0	Gas, atmospheric pressure	39	51,600
			Gas, 6000 psi	16,000	
			Gas, liquefied (−427°F [−255°C])	30,500	
	Town gas[b]	2.0	Gas, atmospheric pressure	16	2,500
			Gas, 3,000 psi	3,200	
	Synthesis gas[c]	2.0	Gas, atmospheric pressure	55	10,000
			Gas, 3,000 psi	7,400	

Exhibit 3.13. Salient Fuel Characteristics (rounded values)

[a]Equivalent volumes; wood and coal values depend on the density of packing. Also, wood and coal have varying energy contents on a weight basis. Burning a pound of coal can release from as little as 4,000 to more than 14,000 Btu. Coal and gas volumetric energies are usually expressed in different units (e.g., Btu/ft³ or Btu/m³); we use Btu/gal here to facilitate comparisons between gas and liquid fuels.

[b]Diluted with atmospheric nitrogen and some combustion products.

[c]SNG

The efficiency of a fuel cell is mostly limited by the "overpotential" and the cell's internal resistance. They result in reducing the voltage at which the current is made available, and hence the energy that can be derived at a given current. The drop in efficiency due to overpotential depends on the nature of the electrode materials used in the cell. The drop due to the cell's internal resistance increases in proportion to the current that is drawn from it. Thus, the energy loss due to the cell's internal resistance is even more severe at the high currents needed for many applications. Transportation is one such application, and the expensive catalysts required for efficient fuel cell operation raise the cost of the cells. Potential uses for fuel cells in transportation include auxiliary power for equipment installed in personal vehicles and for refrigerators in specialty trucks; power for public transport; and eventually power for general vehicle use.

Currently, the only significant use of hydrogen as an energy source—and an extremely small one at that—is in liquid form for rocket propulsion. Its main value in this application is that its combustion provides a high "specific impulse," a critical property for rocket liftoff. This property is a result of the fact that on a per-unit-weight basis, hydrogen has the highest energy density.

Hydrogen can be produced from any hydrocarbon source and is a by-product of some oil-refining operations, with much of that by-product gas used directly in the refining operations. Most commonly, hydrogen is produced by the reaction of natural gas (methane) with steam—a reaction that also produces CO_2. Hydrogen can also be produced by the electrolysis of water. However, there is *no free lunch*: at least as much energy would have to be spent in producing hydrogen by electrolysis as we could ever obtain from the hydrogen itself. What these facts underline is that *water is not a fuel*, and *hydrogen is a not primary energy source*. Rather, hydrogen is an energy carrier, and *could* be an effective storage medium for electricity (see exhibit 3.12), especially when it is coupled with the naturally intermittent photovoltaic and wind sources.

Although using hydrogen in a small fuel cell to supply individual households with electricity (thus avoiding electricity losses of 10% or so in the electrical transmission and distribution system), or as a fuel for automobiles or buses (thus minimizing urban pollution) may seem attractive on efficiency and environmental grounds, there are severe impediments to these uses. The problems center on gaseous hydrogen's very low energy density. Hydrogen's volumetric energy density can be increased by pressurizing it (to, say, 6,000 psi, or 400 atmospheres), but the pressurization process can consume about 10% of the potential energy that would otherwise be available from hydrogen.

Storage and transport of pressurized gas has its own set of challenges. Because storage vessels must withstand high pressures, they must be constructed from heavy steel or other materials bolstered by reinforced fiber, resulting in increased weight for the overall system. Hydrogen can also leak readily through very small holes (at 2.8 times the rate of natural gas). Thus, overall system requirements will be extremely stringent and add to the total weight and cost, thereby detracting from the advantages that the use of hydrogen offers. The other way of increasing hydrogen's volumetric density is by liquefying it. However, the boiling point of hydrogen is very low, which translates into a penalty of as much as 50% of the energy of hydrogen. The minimum energy required for liquefaction is controlled by the same laws that control the conversion of heat to mechanical energy—the lower the required temperature, the more required energy.

Another technology currently under investigation is storage of hydrogen in chemical form as a metal hydride that can be decomposed by heating or by a simple reaction with water. In general, hydrogen accounts for only a small fraction of the hydride's weight. Recently, hydride capture capacity has been improved so that it can store as much as 7% of hydrogen by weight, but that value is still too low for general transportation use. Hydrogen can also be captured by cooled inorganic compounds with complex molecular structures (e.g., zeolites, activated carbon, carbon nanotubes) that release it when heat is applied or containment pressure is lowered. These technologies obviate the need for the very low-temperature (or high-pressure) storage traditionally used, but they may also require cooling to be effective, and they are, in volumetric terms, far less efficient energy-storage media than ethanol or methanol (discussed below). Even as a liquid, the energy content of hydrogen is one-quarter that of gasoline. As previously indicated, but worth repeating, a gallon of gasoline contains more hydrogen than a gallon of liquid hydrogen itself; in other words, gasoline is a more efficient carrier of hydrogen than is hydrogen itself.

The hydrogen used at fixed locations and in some mobile ones can be generated on site, as needed, by processes that convert another fuel to hydrogen. However, such processes require energy and produce CO_2. For example, natural gas under moderate pressure can be delivered through conventional pipelines to stationary locations, where it could be reformed to produce hydrogen for use in fuel cells to produce electricity. The overall thermal efficiency of such a system would depend in large measure on its capacity to use waste heat.

Another challenge facing the widespread adoption of hydrogen has to do with safety. Fire and explosions are a concern whenever a combustible material is used, a feature that hydrogen shares with natural gas and gasoline. However, fires and explosions can occur in air with hydrogen over a substantially larger range of concentrations than with, for instance, gasoline. For this reason, fire and explosion hazards are a real concern with hydrogen. Mitigating this issue is the fact that hydrogen diffuses much more rapidly than does gasoline or compressed natural gas. Yet the media often evoke the image of a fiery crash when reporting hydrogen cars, such as that of the Hindenburg recorded on film in May 1937. For decades this disaster was blamed on a hydrogen explosion and fire. However, later it was shown that fabric coatings within the vehicle were the major fuel for the fire. This is just an example of how the danger from hydrogen, while real, has been overhyped.

Introduction of hydrogen in commercial and residential applications will have to overcome concerns with the safety of the physical plant. New

---

"CO$_2$-FREE" ENERGY: A CAUTION

Many suggest that a major advantage of hydrogen as a fuel is that it not only is clean burning but also emits no CO_2. However, that is true from a global perspective only if hydrogen's manufacture is itself CO_2-free, that is, if it is produced from energy derived only from nuclear power or solar technologies, or to the extent that the CO_2 generated in its production using other sources is captured by growing biomass.

regulations will certainly be imposed, and formulation of acceptable state or nationwide safety and handling standards will undoubtedly require many years to accomplish.

Ethanol

The push for ethanol from biomass to serve as a gasoline extender in automobiles has resulted largely from concerns about energy security as nations try to minimize their dependence on imported oil. Ethanol can be produced in petrochemical plants from hydrocarbons containing two or more carbon atoms. It is also produced (more expensively today) by fermentation of the natural sugars in crops such as corn and sugar cane, as well as by enzymatic hydrolysis and fermentation of biomass materials such as wood, corn stover (stalks and leaves), bagasse (sugar cane residues), and rapidly growing reeds and grasses.

Ethanol production is receiving government subsidies of various kinds to achieve social purposes (e.g., support of an agricultural community in the United States) or to avoid the balance of payments deficit created by oil importation (e.g., as in Brazil). In 2006, Brazil and the United States each produced around 7.0 billion gallons of ethanol, and production is slated to grow substantially in both countries. However, ethanol's contribution to overall global energy is currently insignificant. We discuss the production of ethanol and other biofuels in detail in chapter 7.

Methanol

Methanol was once produced by the "destructive" distillation of wood. Today its primary source is synthesis gas (a mixture of CO and hydrogen) generally created by chemical reactions involving natural gas or other hydrocarbon sources, including biomass. Natural gas obtained from wells in remote locations is also sometimes converted to methanol. Producing methanol from natural gas results in a conversion energy loss of about 30%, but liquid methanol is easier and safer to ship to points of use than is liquefied natural gas. Various other thermal processes that use biomass as a feedstock are also available, and gasoline can be produced from woody material using methanol as an intermediate. A major problem with methanol's use is that its volumetric energy density is about 45% that of gasoline, and its combustion can produce small amounts of formaldehyde, which causes eye irritation among other harmful effects. Methanol could also be used as a fuel in certain direct methanol fuel cells to produce electricity; significant progress has been made in developing this technology, but the process is not ready for use in large-scale applications and is still too expensive.

Low-Btu Gases

Gases made by the partial combustion of biomass, coal, and oil also constitute secondary energy sources. These are mixtures of CO and hydrogen. The simplest, "town gas," is made by the partial combustion of wood, coal, or oil in air to produce a gas with an energy content of about 120 Btu/scf and was used in the mid to late 1800s and even after for street lighting in towns (hence the name) until it was replaced by electricity. If oxygen is substituted for air, the result is a "medium-Btu gas" product, which is a mixture of carbon monoxide (CO), hydrogen (H_2), carbon dioxide (CO_2), and steam (H_2O), which is also referred to as synthesis gas or syn gas. Only the first two of the constituents, CO and H_2, are combustible, and the product has an energy content ranging from 450 to 600 Btu/scf. Further chemical manipulation of this product produces what is called "substitute natural gas" (SNG) with a heating content of about 1,000 Btu/scf; that energy content is typical for natural gas, so SNG can be used in all applications that are now served by natural gas.

Summary
Secondary fuels require energy and additional investments to produce. Therefore, no matter what their origin, they are likely to remain significantly more expensive than oil until the depletion of conventional oil resources necessitates use of more expensive methods of recovering oil from unconventional fossil resources. Secondary fuels are also more difficult to use because of their lower volumetric energy density. They could eventually play a major role in the transportation field, but substantial adoption by a vehicle fleet could take decades. Hydrogen produced from water by solar or nuclear electricity is not likely to play an important role for several decades because logistic concerns require significant investments in infrastructure compared with other methods that could achieve the goals of reducing our dependence on oil or reducing greenhouse gas emissions.

A Look Ahead

The energy industry and its customers are engaged in a massive and complex venture that has been, and will continue to be, in flux. Eventually, new energy sources that first supplement and then replace fossil fuels will be required to satisfy increasing demand for energy. In many cases, the new sources of energy will probably be both harder to use and more expensive to produce. Thus, their adoption may prove difficult to achieve quickly.

A critical question facing the world today is: *How large is our remaining "assured" global oil resource base?*[16] The actual amounts can be debated, but as discussed in chapter 5, the assured conventional oil resources are unlikely to be sufficient for much more than a few decades. Even speculative conventional oil resources may no longer be available much beyond mid-century—especially

---

16. In discussing fossil and mineral resources throughout the book, "assured" quantities are those that professionals judge to have at least a 95% chance of recovery, "likely" quantities are judged to have a 50% (or statistical median—usually close to 50%) chance, and "speculative" quantities are judged to have only a 5% chance of full extraction of the quantity stated.

if oil use continues to increase at its current rate. Development of improved technologies for tapping into unconventional resources such as tar sands and oil shale will be needed to avert that shortage. Even more important, battles for control of these resources could well begin long before we approach "the last drop." Similar statements apply to natural gas, although "crunch time" for that resource will likely come later than for oil. Current annual gas use is only three-fifths that of oil, whereas assured gas reserves are about the same as those for oil, as discussed in chapter 5.

Although coal resources are substantially larger and could last for more than a century, they are less desirable because their use entails even more greenhouse gas emissions, unless technologies for capturing and sequestering CO_2 are used. Currently, those CO_2-abatement technologies are in the development stage and are expensive, and their long-term effectiveness is unknown. The scale at which they can be used in the next generation of coal-fired power plants is under debate. Uncertainties regarding different energy sources are discussed throughout the book, as is the potential for reducing, perhaps even reversing, growth in energy demand. These uncertainties also raise several key questions, all of which the public and policy makers must address:

- Will the social benefits and more benign environmental aspects of wind and solar energy sources outweigh the potential disadvantages associated with their higher cost and potential lack of availability at the times they are needed?
- How will our energy system—the energy industry, its suppliers, and its billions of customers—be affected by continually increasing demands for environmentally cleaner energy?
- How can we install any new technology at a truly significant scale?
- Will we adopt the newer technologies because they are cleaner? Or will we be forced to use the newer sources despite their drawbacks because we have so depleted the older and easier-to-use resources that their extraction has become too costly?

Subsequent chapters explore further details of the global energy system. As you read along, we urge you to consider what energy future(s) would be best for us, our offspring, and the world over the coming years. You should also consider how the decisions necessary to achieve that future can be made and implemented. Some of our ideas about these matters are presented in chapter 9.

4

Energy Needs to 2050

Having looked at the evolution of energy use and the current state of the energy industry in the previous chapters, we are now ready to make some projections for the future. As the famous Danish physicist Niels Bohr once said, "Prediction is very difficult, especially about the future."[1] Forecasts of human activity are fraught with uncertainty. This is clearly true of energy forecasts, given that regional and national economic, political, and social trends can change world energy use, as can scientific discoveries and engineering developments unanticipated when the forecast was made.

Among the technological changes that have had the largest impact on our pattern of energy consumption, perhaps foremost is the development of the internal combustion engine in transportation, and the accompanying enormous increase in the use of petroleum. The mobility afforded to individuals by automobiles, trains, and later by airplanes greatly shrunk the world, bringing people and economies closer together but also in many cases, unfortunately, helping precipitate international conflicts. Perhaps equally important has been the rise in the use of electricity. The development of the electric power system and its extension beyond urban areas profoundly influenced the daily lives of everyone touched by it. Likewise, the discovery of semiconductors and their application in information technologies, including computers, the Internet, wireless personal communication, and space-based global

---

1. Yogi Berra is also quoted to have made a similar observation: "It's tough to make predictions, especially about the future."

communication, have altered the way we interact with one another and our surroundings. While the use of these technologies has increased the demand for energy—particularly electricity—it has also contributed immensely to our productivity and thus at the same time helped mitigate the increase.

The discovery of atomic fission and its uses in peace and in war have had ramifications in the technological and geopolitical realm. Following the Second World War, great optimism prevailed as exemplified by the statement in 1954 attributed to Lewis Strauss, then chairman of the U.S. Atomic Energy Commission, that nuclear power will be "too cheap to meter."[2] The ability of nuclear power to deliver electricity abundantly and cheaply helped it become a significant contributor to global primary energy within 20 years. However, much of the expansion was driven by concerns about political hegemony and national pride, and not about supplying power. After the initial impetus, nuclear power essentially ceased to grow in the 1980s in the United States, and markedly slowed down in the 1990s worldwide. There are several reasons that the growth of the nuclear industry slowed and stopped. Foremost was the emotional impact of two nuclear accidents, at Three Mile Island in 1979 and at Chernobyl in 1986. The other reasons have to do with hard economics. The cost of nuclear plants did not decrease as anticipated, and the demand for electricity did not grow as projected as a result of the increased conservation and efficiency measures that were adopted in the wake of oil embargoes of the 1970s. Taken together, these reasons did away with much of the incentive for building new plants.

In the political arena, in addition to the oil embargo just mentioned, several unforeseen events have had major impacts on the global energy picture. The breakup of the Soviet Union into independent states altered global attitudes and geopolitics, with attendant changes in national and international policies related to the development of nuclear weapons and nuclear energy programs. China has embraced the market economy and transformed into an economic powerhouse that wields international influence, and one that requires substantial amounts of imported energy. In the 1980s, China was projecting rapid increases in domestic oil production leading to huge surpluses. Those increases in production did not materialize, but China's demand for oil has dramatically increased. It is now the third largest oil importer, after the United States and Japan. China has planned a substantial nuclear power program to supplement, and exceed, its large commitment to hydropower.

The globalization and technological advances of the last two decades have meant that large numbers of people throughout the world, particularly in developing nations such as China, India, and Brazil, are now empowered to enter the global market with full participation. We do not mean to suggest that the prosperity brought about by globalization has uniformly helped everyone in the

---

2. The statement has been interpreted to mean that electricity would be free, although Strauss was probably referring to a business model wherein the electric service would provided at a flat rate regardless of usage.

developing nations. In many instances, the conditions for large segments of populations within these countries have worsened as a result of losing their means of livelihood, even as the standard of living has risen for a different group within the same country. Given both the increased demand and the heightened need for globalized markets, several key questions arise regarding the demand and supply of energy. Will the future bring comparable game-changing events with impacts as great as these recent developments have had on our energy economy?

- How will global energy needs be affected if the United States suffers a protracted, severe economic recession with the attendant severe repercussions to the highly intertwined global economy that we now live in? Since this book was first drafted, the United States—and perhaps the entire world—has faced a significant downturn in economic well-being and a consequent lower demand for energy.
- What happens to global energy supply if Saudi Arabia is thrown into a state of anarchy following the demise of its king? Will the absence of a clearly identified successor lead to armed disputes between the ultraconservative and more moderate religious groups? Will the petroleum-rich countries of the Middle East and the industrialized and emerging economies engage in a fierce competition for the energy resources of the entire Middle East region, accompanied by untold human suffering from war and the consequent disruptions in energy supply?
- Will fears of global warming cause many nations to engage in massive efforts to reduce CO_2 emissions from fossil fuel combustion by imposing a steep price on carbon emissions? Such actions would raise the price of energy and likely lead to a substantial reduction in energy use, but they might also deny billions of people, who are currently consuming the least energy and are least able to afford the higher prices, an opportunity to improve their standard of living.
- What happens to energy production and demand if a major epidemic defeats modern medicine and kills tens or even hundreds of millions, causing the world's population and productivity to decline substantially rather than increasing as most anticipate?
- What will be the impact if a simple, uncomplicated and inexpensive way to create and harness fusion energy for electric power generation finally emerges?
- How does the overall energy scenario change if a large-scale, low-energy-loss, low-cost electrical storage system is developed, and/or we find much more effective means to produce, store, transport, and use hydrogen as a fuel?

Long-range prediction models rarely take into account the dramatic changes that could ensue should any of the above examples come to pass. Recently, two retrospective studies examined long-range forecasts of U.S. energy use made since about 1950 to identify, with the benefit of hindsight, those factors that forecasters have consistently predicted accurately and those that they have consistently been wrong about.[3] Both studies revealed that forecasters have tended to underestimate the extent of energy resources, particularly of oil and gas. They have also underestimated the role of market economics and have predicted far greater use of energy than has been borne out by history: "The free market works, often with a vengeance, but this seems to be a lesson *not* learned." There is also a tendency to embrace remarkable advances in some new technology that is often "hyped" at the peril of overemphasizing its role in the future. Finally, according to the authors of the retrospective studies, forecasters tend to display a lack of appreciation of the mobility of targets and improvement of technologies with time.

These retrospective studies also note the influence of funding organizations and their particular objectives on the results. An organization will have bias to fund studies in areas that best respond to its particular interest—while ignoring others that might have at least a peripheral influence on the study result, whether they be about the growth of a particular market or a policy objective.[4] In its forecast, the Gas Research Institute concentrated on gas and promoted its potential. A collaboration of five U.S. Department of Energy laboratories (cochaired by a former member of the SRI staff) concentrated on the need for U.S. government research, and the American Council for Energy Efficient Economy emphasized the importance of efficiency in both energy use and renewable technology in its forecast. Similarly, Shell analysts have consistently ignored the potential of nuclear power and emphasized the potential of new vehicle technology in their global scenario technology reports. We believe that despite public fears, nuclear energy will play an increasingly important role in providing energy in the future.

Predictions of energy futures are not facts; no matter how carefully crafted, they are primarily guesses about the future. Yet limited as they may be, they are better than nothing. As University of California–Davis energy policy expert Paul Craig has noted in writing for the Sierra Club, a compilation of such predictions can help us place bounds upon our true energy futures. That is the spirit in which we approach our predictions about future energy

3. P. P. Craig, A. Gadgil, and J. G. Koomey. "What can history teach us? A retrospective examination of long-term energy forecasts for the United States." *Annual Review of Energy and the Environment,* vol. 27, 2002, 83–118; R. H. Bezdek and R. M. Wendling. "A half century of long-range energy forecasts: errors made, lessons learned, and implications for forecasting." *Journal of Fusion Energy,* vol. 21, 2002, 155–172.

4. We, the authors of this book, admit that we are not free of influence either. E.M.K. was long associated with nuclear affairs beginning as early as 1943, and he studied nuclear proliferation issues in the 1960s and energy alternatives in the 1980s and 1990s. H.D.C. had developed his practical interest in conservation efforts beginning in the 1970s. R.M. has been actively conducting research on advanced coal conversion processes for almost 30 years and since the 1990s has been increasingly engaged in research on bioenergy. We hope that these varied backgrounds, combined with a desire to present a global approach to the use of many energy options, will act to reduce such biases.

Exhibit 4.1.
Key 2006 Economic Statistics for the 10 Most Populous Countries

Country	Population (millions)	Energy Use per Capita (GO)	Per Capita GDP (dollars)	Growth Rate in GDP (percent)
China	1,320	361	5,345	11.9%
India	1,123	105	2,753	9.0%
United States	302	2,174	45,793	2.2%
Indonesia	226	137	3,728	6.3%
Brazil	192	306	9,570	5.4%
Pakistan	162	99	2,524	6.4%
Bangladesh	159	35	1,242	6.5%
Nigeria	148	1,403	1,977	6.3%
Russian Federation	142	1,149	14,747	8.1%
Japan	128	399	33,517	2.1%
World	**6612**	**497**	**9,896**	**3.8%**

demand in this book. We analyze global energy requirements under a range of outcomes encompassing various possibilities.

Whether or when game-changing events like those cited above occur are unknown, and such events are generally not included in forecasting. In their absence, there is little argument about the ongoing rise in demand for energy. The per capita gross world product (GWP) was just under $10,000 (purchasing power parity) in 2007. That average is well above the subsistence level and could be considered a conservative measure of income, above which demands for goods and services—and thus energy—will rapidly increase. Exhibit 4.1 lists several key growth and energy statistics for the 10 most populous countries that together account for 63% of the world's population. The data for population, per capita energy consumption, GDP, and recent GDP growth rates, as well as the comparable figures for the world as a whole, are from the World Bank's *World Development Indicators.*[5] The per capita GDP in China was $5,345 in 2006 and growing at an annual rate of 11.9%. At that rate, China's per capita GDP would exceed the $10,000 mark in 6years. India's somewhat slower growing economy could take until 2024 to pass the $10,000 mark. These prospects argue for a substantial increase in global energy demand over the next decade or so. Although the economic downturn in the second half of 2008 has the potential to slow down future economic growth, recent indications are that the slow down will not be as severe in the emerging economies, where most of the increase in energy demand is expected to occur.[6]

5. World Bank. *World Development Indicators,* revised October 17, 2008. http://go.worldbank.org/SDHOTB92H0. Accessed November 2008.

6. Preliminary data suggest that China's growth in 2008 slowed to a "mere" 8.9%, still indicating a rapid pace.

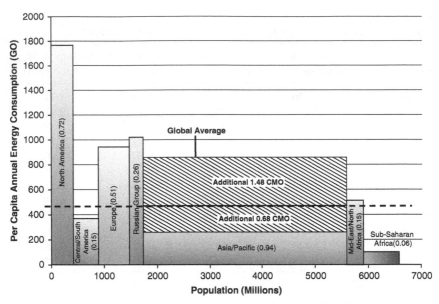

Exhibit 4.2. Regional Per Capita Energy Consumption and Population in 2006

The drivers for increased energy demand are population growth and rising standards of living. Currently, large segments of the world's population consume energy at less than a third of the global average. They are striving hard to better their economic condition, and globalization is facilitating their ability to become fuller participants in the world economy. As countries advance toward that goal, their energy consumption is bound to rise. Exhibit 4.2 allows us to get more quantitative about the projected increases—it duplicates exhibit 3.8 but with the addition of two rectangles showing the impact that an increase in the Asia/Pacific region's per capita energy consumption would have on total energy demand (hatched area). The lower rectangle shows an increase in energy demand of 0.68 CMO if the per capita consumption in the Asia/Pacific increases to the current global average of 450 GO. The upper rectangle shows the impact of increasing consumption in the Asia/Pacific region to half that of the North America (not just the U.S.) regional average of 1,800 GO (i.e., 900 GO). Note that at 900 GO, per capita energy consumption in the Asia/Pacific region will still be lower than current consumption in Europe (970 GO). The amount of additional energy needed in the two cases would be 0.7 and 1.3 CMO/yr, respectively.[7] As mentioned above, in view of the economic growth in China and India, these growth scenarios are not unlikely unless they are slowed by energy shortages. This simple exercise highlights the point that strong drivers

---

7. Hatched areas show the additional energy needed if the Asia/Pacific region increases its consumption to (a) the global average, and (b) half that of the North America region.

are already in place for an increase in annual energy demand by about 1 CMO or more in the next few decades.

While increased energy demand is almost assured, increasing energy efficiency is a countervailing driver that can reduce energy consumption. Increases in energy efficiency have been steadily underway for a long time. The history of U.S. energy consumption shows that energy efficiency has generally increased over extended periods except for some decline in the 1895–1930 period.[8] We have every reason to expect that continued innovation would allow us to further check the growth of, or even reduce, the per capita energy demand. We discuss the growth in energy efficiency—which is the monetary value of output per unit of energy consumed—and its potential impact on energy consumption in greater detail in chapter 8.

Innovations in energy efficiency can potentially reduce energy demand, but for innovations to have a significant impact on curbing the projected rise in energy demand, they must provide reductions in energy use on the scale of a CMO, because that is the scale of the projected increases in global energy demand. Large savings of energy require that such innovations be adopted widely. And for widespread adoption to occur, the innovations must cost no more than the amounts that large populations are willing to pay. As we discuss in chapter 8, many of the energy-saving techniques in use today and antici-pated for the future deliver *far* short of the desired CMO-scale reduction.

Many forecasters, including those at the U.S. Energy Information Agency (EIA), use sophisticated approaches to project energy use. These approaches use economic models that predict growth (or decline) in GDP and then mul-tiply that value by an "energy intensity" factor. This factor is the amount of energy needed to produce a unit of GDP; energy intensity is the inverse of energy efficiency, which measures the GDP output per unit of energy con-sumed. These models take into account historical growth rates of GDP, expected trends in energy intensity, and population. These factors are interdependent and can either amplify or attenuate each other. For example, increased wealth has historically acted as a strong negative influence on population growth, and thus total energy consumption resulting from increased GDP would be less than predicted by a simple multiplication of GDP and energy intensity. The calculus is somewhat involved, but no matter how complicated, it always comes down to making intelligent guesses about the rates of changes in GDP, population, and energy intensity. These guesses are often extrapolations of recent trends and do not anticipate potential disruptions such as an outbreak of a war or a major disease epidemic that would radically alter the pattern of energy use. Therefore, predictions of energy use made by this approach ordi-narily do to not vary by the larger factors seen at times of crisis.

---

8. The decline resulted in large part from the rapid increase in the use of commercial energy, including the introduction of then-inefficient electricity generators and of inefficient internal combustion engines in the rapidly growing transportation sector.

Scenario building as used by analysts at the Shell Energy Group and a number of other public and private institutions is another approach to forecasting. This approach anticipates disruptions that may occur in the political, economic, or technological spheres and constructs various possible outcomes or scenarios. The energy use profile for each of these scenarios is then estimated from the different GDP, population, and energy intensity assumptions that make up the particular scenario.

Shell indicates that its results are *possible outcomes,* not certainties. Shell's projections show average growth rates ranging from a low of about 0.8%/yr to a high of 2.6%/yr. The EIA, which tends to be conservative and in the past has been the most accurate in predicting future energy use, has recently published its forecasts to 2030. The EIA projects energy use to increase by 1.8%/yr (the reference case). In addition, the EIA has calculated high-growth and a low-growth cases under which energy use growth is 2.1 and 1.4%/yr, respectively.

To analyze global energy needs through 2050, we have chosen four cases that span the entire range of *possible* outcomes. Exhibit 4.3 depicts annual energy use resulting in the four cases, and the main points are also tabulated in exhibit 4.4. For the high-energy-use case, we have used a business-as-usual scenario based on the world's consumption of primary energy since 1965. Global energy use increased steadily from 1.07 CMO in 1965 to 3.01 CMO in 2006, which corresponds to an average growth of 2.6%/yr. Were this rate of growth to continue, the global energy demand in 2050 would be 9.4 CMO. This case corresponds to Shell's high-growth scenario. For the low end, we have chosen a rate of 0.8%/yr, which was the projection of the World Energy Council (WEC) and the low-growth scenario of Shell. This low growth rate results in annual consumption of 3.9 CMO in 2050.

We also include our extrapolation to 2050 of the EIA reference predictions (1.8%/yr), which were made only to 2030, as intermediate values. The EIA's reference case, as well as its high- and low-growth predictions, fall between the high and low values we use herein. Under the intermediate value, global energy consumption would climb to 6.4 CMO in 2050. Under the EIA's high- and low-growth cases, global consumption would be 7.3 and 5.2 CMO, respectively, in 2050.[9] It is noteworthy that the world's consumption for 2006 already surpassed the predictions that most of these entities made between 1996 and 2001 and appears to more closely match the highest growth rate of 2.6%/yr.

Exhibit 4.3 also includes a fourth case in which the current high growth rate (2.6%/yr) is in effect initially but then slows as the world undertakes highly stringent measures to increase energy conservation and efficiency. Under this

9. In their more recent forecasts, EIA and the Shell Corporation, among others, have revised their energy growth projections downward such that extrapolating them would place global annual energy consumption in 2050 at around 6.5 CMO.

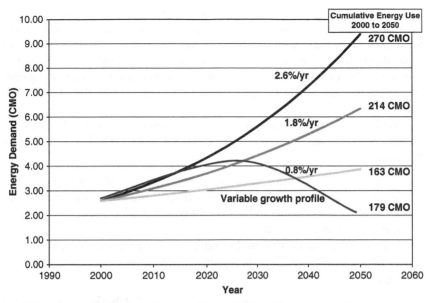

Exhibit 4.3. Predicted Global Energy Use under Four Cases

Exhibit 4.4.
Assumed Growth Rates and Predicted Energy Use for Four Cases

Case	Assumed Growth Rate (%/yr)	Predicted Energy Use in 2050 (CMO)	Cumulative Energy Use Between 2000 and 2050 (CMO)
1. Business as Usual (High Growth)	2.6	9.7	270
2. EIA Reference Case	1.8	6.4	214
3. WEC (Low Growth)	0.8	3.9	163
4. Mixed Growth	Variable	2.0	179

case, the growth rate reverses after 2030 such that by 2050 total energy use is 80% of that in 1980, or about 2.0 CMO/yr. Such a scenario might be required if we are to prevent atmospheric CO_2 concentration from exceeding 550 ppm and thus limiting the projected global warming to less than 3°C.

Under these different cases, the world's cumulative energy use between 2000 and 2050 would amount to about 270, 214, 163, and 179 CMO for the business-as-usual, EIA reference, low-growth, and mixed-growth scenarios, respectively. These sums are very much larger than the estimated 93 CMO that was consumed in the 20th century. If the rise in global energy use continues unabated at the rate of the past century (business-as-usual scenario), we will use almost three times as much energy in the 50 years between 2000 and 2050 as we did in the previous 150 years. Even under the mixed-growth scenario, case 4, in which energy use reaches a peak and then declines, the total global energy use over the 2000–2050 period would be almost twice that of the entire last century!

Which of these cases is more likely to be closer to the actual outcome? We do not know. Nevertheless, we can draw two major conclusions from the projections:

1. There will be continuing pressure on supply at least through 2050 as the world energy use keeps growing at significant rates. If this were to occur we could argue whether in the next 50 years the Malthusian resource limitation will be upon us.
2. Global energy consumption through 2050 is expected to exceed that of the entire past 150 years by significant amounts, even if the current rapid growth is reversed.

These are not unreasonable expectations. Almost 2.4 billion people in the developing world, or more than one-third of world's total population, live in countries whose average per capita income is approaching the aforementioned $10,000 mark, a level often viewed as the "trigger" point for substantial growth in the use of consumer goods. Even the mixed growth scenario, the one with lowest projected energy use in 2050, anticipates a substantial global effort to reduce energy consumption, but one that still assumes overall growth for at least the next two decades. Will the ultimate low energy consumption come to pass because the world begins to practice efficiency and conservation earnestly, or simply as the result of the energy industry being unable to meet the market demands at an acceptable price? The former is preferable because the latter possibility does not bode well for peace and security among nations.

Challenges Ahead

While producing the additional energy required in 2050 from fossil fuel is not outside the realm of possibility, it would run up against the world's ability to produce oil at the required rate. Moreover, as global awareness of the cumulative environmental effects of fossil fuel use rises, resistance to its use is likely to increase. If we decide to curtail or eliminate the use of fossil fuels, the global energy system will have to change radically, either by a substantial reduction in the energy demand or by rapid development of large-scale solar thermal, wind or nuclear power. *These are the only technologies with sufficient resources for lasting increases in energy supply.* And, scaling any of them to a CMO/yr level is a daunting task. Chapter 6 discusses the potential and challenges for nuclear power production and use, and chapter 7 describes the same for solar and other income energy sources.

The transition to new forms of energy use will be a fundamental one, and not without painful ramifications. A challenge during the next few decades will be to reverse the continuing increase in fossil fuel use while supplying a major part of an ever larger total energy requirement from alternative sources. Because oil and other fossil resources are moving toward depletion, the world will ultimately be forced to live on "income" from solar or nuclear technologies to supply the much larger total and per capita quantities of energy that we will need.

Another challenge will be learning to shift from the relatively profligate energy-consuming lifestyle in the developed world, which is drawing with increasing rapidity on our declining inherited resources, to one that lives comfortably within the limitation of its energy income. From this perspective, uncertainty regarding the amount of fossil fuel still present in an economically recoverable form clouds our planning (this is discussed in chapter 5). The developed nations will need to limit their demand for energy-intensive goods and services, yet the total energy savings possible by reducing consumption in the developed countries is insufficient to offset the increased demand that the developing nations will soon be placing. Consider the fact that the *total* energy use in the North America region in 2006 was 0.7 CMO. If half of that were saved, a highly unlikely scenario, the net savings would be 0.35 CMO while the projected annual energy demand worldwide is likely to increase by more than 4 CMO by 2050! Clearly, conservation by nations considered fully developed right now will be inadequate to singularly solve the energy issues of 2050.

As another example, consider the world's automobile and truck fleet, now consisting of more than 500 million petroleum-consuming vehicles. This fleet might well increase by 25–50% within the next few decades, with most of the additional vehicles sold in the developing world. Even if these new vehicles are not gas-guzzlers, the demand for oil will continue to increase rapidly. Higher oil prices may slow growth in consumption, but only at the expense of pent-up demand and global unrest.

Although a move to solar-driven income sources may become possible sometime in the future, a large shift to income energy sources before 2050 seems extremely unlikely because of complex technical, economic, and infrastructural issues of production and distribution. We cannot simply turn off the fossil fuel systems and replace them with income resources without causing a great imbalance in the energy supply and demand. Nuclear power could help bridge this transition away from CO_2-emitting energy sources. It might be gradually abandoned when, and if, solar energy and conservation efforts become truly effective and/or nuclear fusion proves to be a practical source of energy. It is our belief that the best we can achieve between now and 2050 is to move *toward* such a state with actions carefully chosen and without excessive delay so that we can avoid major economic, political, and social dislocations.

Minimizing the possibility of a turbulent transition will require us to address the supply and demand sides of the energy balance simultaneously. We will need to reduce the present rate of growth in energy demand, and simultaneously initiate a large energy supply program that includes varied forms of solar-based energy as well as a vigorous nuclear component, which we believe will be needed for a considerable time.

Part II

Energy Resources

Part II

5

Our Energy Inheritance

Fossil Fuels

The use of fossil fuels—petroleum, natural gas, and coal—is ubiquitous today and has made possible the advances of modern civilization. These fuels are capable of providing energy for a variety of applications—from very small to very large—and touch our lives in many ways. A small gas-fired heater uses about 50,000 Btu/hr (1 standard cubic foot [scf] of gas/min) and keeps our homes warm. A 200-horsepower gasoline engine in a family car consumes around 2 gal/hr of oil and can carry a load of five passengers a distance of 60 miles on a level highway. An 1,800-ton/hr cement plant consumes 900 MBtu/hr (about 0.9 million scf gas/hr) when in full operation and produces the building material widely used for constructing homes, offices, industries, roads, and bridges. A large, coal-fired electric power station (1,000 MW rating) requires between 300 and 500 tons of coal per hour and produces enough electricity to power half a million homes.

The range of power that fossil fuels, particularly oil, can deliver is truly amazing: the same basic fuel that powers jet aircraft also powers children's model aircraft engines. It is unlikely that aircraft will ever be powered by solar panels mounted on the wings or by on-board nuclear reactors. The importance of fossil fuels in our lives cannot be overemphasized. It took millions of years to accumulate them, and their potential exhaustion in just a few centuries should seriously concern all of us.

In this chapter, we briefly review the circumstances that led to formation of our fossil fuels and then discuss how much of each of them is available. This discussion requires clarifying the special meanings ascribed to such terms as reserves and resources. For all three fuels, we look at the global distribution of our resources. We also present estimates of possible resource lifetimes under varying conditions of use and indicate the nominal equipment and infrastructure requirements for producing these inherited resources at a rate of 1 CMO/yr. As we shall see, our conventional reserves are somewhat limited, but our resource base is large, and unconventional oil and gas resources offer a substantially greater potential. Nonetheless, exploiting unconventional resources is certain to be more expensive and, in most cases, potentially more damaging to the environment. Thus, it boils down to making choices about which resource(s) to develop.

Formation of Fossil Resources

Petroleum most often refers to a liquid consisting of a complex mixture of hydrocarbon molecules, each composed of anywhere from 2 to 40 carbon atoms arranged in a linear, branched, or ring molecular structure. Occasionally, these molecules may also contain sulfur, oxygen, and nitrogen atoms, as well as smaller amounts of nickel and vanadium. The term "petroleum" encompasses gases associated with oils, heavy oil or bitumen, and even the oil precursor, kerogen. Natural gas is part of the petroleum family and is composed of smaller molecules, mostly methane—which has one carbon atom— and also of molecules containing two to five carbon atoms. Therefore, natural gas often occurs along with petroleum, although there are also many instances when natural gas occurs by itself. Coal is also a complex mixture of hydrocarbon molecules, mostly with ring structures that are often cross-linked to one another to form a network that is solid under ambient conditions. Coal also contains varying amounts of oxygen, sulfur, and nitrogen atoms. Very often it is associated with minerals containing toxic elements such as mercury, arsenic, and even some uranium.

Petroleum was formed millions of years ago from the debris of microorganisms, mainly plankton algae, after they fell to the ocean floor and mixed with mud to form the petroleum source rock. The source rock was then drawn down by geological forces—*subducted*—deep into Earth. There it was essentially cooked under pressure in the absence of oxygen, producing the hydrocarbon molecules that make up crude oil. Geologists refer to the depth over which petroleum is formed as an "oil window." If the subduction went deeper than around 16,000 ft (~5,000 m), where the temperature exceeds 480°F (250°C), natural gas rather than oil became the main product. The new "Jack" oil field discovered in Gulf of Mexico waters at about 26,000 ft deep is evidently an exception to this general rule. In this case the oil may have resulted from either a short burial or a decreased local geothermal gradient.

The resulting hydrocarbon species then migrated upward through the rocks toward Earth's surface. At the surface, natural gas and the lighter hydrocarbons were released to the atmosphere—sometimes resulting in the "burning pillars" mentioned by ancient scribes. The higher boiling materials may have accumulated in tar pits, some of which are still around.[1] Some fraction of the upward-migrating hydrocarbons was trapped below impervious rock formations, the places where we generally find crude oil deposits today. Oil and gas are not found in giant pools as many pictures in textbooks and the press suggest, but are instead trapped in small voids between various sands and rocks. If the petroleum accumulation does not have a cap of impervious rock, then its more volatile portions slowly evaporate, and the result after extended periods is tar sands such as those found in Alberta, Canada. On the other hand, if the kerogen precursor to petroleum is not subjected to high enough temperatures to form oil, then oil shale—essentially oil precursors in rock form—results. Oil can be produced from the shale with additional thermal treatment.

An opposing "abiotic" theory posits that all petroleum is primordial, formed from the reaction of accreted carbon with water, deep within Earth.[2] During the migration of petroleum to near the surface, it picked up microfossils and biomarker molecules that—according to the proponents of this theory—have been incorrectly cited as evidence for the biotic origin of petroleum. Adherents to this minority point of view assert that Earth has an extremely large, virtually unlimited, quantity of petroleum that will refill the reservoirs. While reservoir depletion is a common phenomenon, there are no convincing examples of depleted reservoirs getting refilled, at least not on any timescale relevant to the current human civilization. The proponents further assert that, led by the incorrect biotic theory, geologists have been looking for oil in the wrong places. However, so far only one drilling expedition was based on the "abiotic" theory, at Siljan Ring in Sweden. That experiment was a resounding failure. Of course, this failure does not mean that hydrocarbons cannot be formed by abiotic processes. They certainly can, and are.[3] But whether those processes are indeed what produced the petroleum that we use is another matter. And while theories about oil's origin might help guide future prospecting, for example, by drawing attention to meteor impact craters like the Siljan Ring or where plate tectonics have formed deep fissures, oil explorers have already looked at almost all accessible regions of the globe. Until and unless new fields

1. The La Brea Tar Pits in central Los Angeles is an example. Judging from the bones of saber-tooth tigers and other fauna that have been found in them, the pits date back at least 10,000 years.

2. J. F. Kenney, V. A. Kutcherov, N. A. Bendeliani, and V. A. Alekseev. "The evolution of multicomponent systems at high pressures: VI. The thermodynamic stability of the hydrogen-carbon system: the genesis of hydrocarbons and the origin of petroleum," *Proceedings of the National Academy of Sciences of the United States of America*, vol. 99, pp. 10976–10981, 2002.

3. H. P. Scott, R. J. Hemley, H. K. Mao, D. R. Herschbach, L. E. Fried, W. M. Howard, and S. Bastea. "Generation of methane in the earth's mantle: in situ high pressure-temperature measurements of carbonate reduction," *Proceedings of the National Academy of Sciences of the United States of America*, vol. 101, pp. 14023–10426, 2004.

are found, our reserve and resource levels will remain relatively unchanged, only decreased by the amounts consumed. These reserve and resource quantities are not affected by *how* oil was originally formed. Rather, these estimated quantities are determined first by whether we find oil, and second by how hard is it to extract the oil.

Around the globe, we have more than 40,000 known oil fields that contain about 850,000 producing wellheads. Most of the fields are on land, but more and more are being developed offshore at water depths approaching 8,000 ft (2,500 m). The fields are also of varying sizes, with a few stretching for many tens of miles. By drawing on their knowledge of the subsurface rock structure and techniques such as three-dimensional seismic imaging, petroleum engineers can now locate probable underground oil accumulations, or fields, from which oil can be extracted by drilling. In examining geological strata for oil, the time of interest stretches from the pre-Cambrian age of 600 million years through the Triassic period of about 200 million years ago. In many oil fields, the oil is under pressure and easily flows out of the wellhead, along with associated gas and water. In others, it is necessary to pump the oil from the well or to drill additional holes to force down water or CO_2 to cause the upward flow of oil. How readily the oil flows out (i.e., oil field "productivity") depends on many factors; principal among them are the structure of the rock in which the oil is trapped, especially the rock's degree of porosity; the viscosity of the oil; and the presence or absence of a gas dome or water reservoir that supplies pressure to the field. Producing oil from a field requires careful management of well pressure to prevent the collapse of rock porosity; such a collapse would choke off the oil flow or result in undesirable flow of water and gas to the surface along with the oil.

Oil fields vary greatly in productivity, ranging from a few barrels to more than several hundred thousand barrels a day, for a current worldwide total of about 90 million barrels every day. The world's largest and most productive oil field is Ghawar in Saudi Arabia. About 174 miles long with a maximum width of 19 miles, it produces about 5 million bbl/day of oil and has been doing so for more than 30 years. At present the high output is maintained by injection of a quantity of water similar to that of the oil produced. Output from Ghawar represents more than half of Saudi Arabia's total production. The next five most productive fields in that country each produce between 400,000 and 1 million bbl/day. Given the age of these oil fields, however, and the difficulties of maintaining oil reservoir pressures, Matthew Simmons, founder and CEO of Simmons & Co. International, an investment banking company specializing in energy, has pointed out that Saudi Arabia may not be able to sustain overall oil production of 10 million bbl/day,[4] let alone achieve the 14–16 million

4. Matthew R. Simmons. *Twilight in the Desert: The Coming Saudi Oil Shock and the World Economy* John Wiley and Sons, New York, 2005.

bbl/day called for in U.S. Energy Information Agency (EIA) projections to meet world energy demands to 2030.[5]

Coal, our other major fossil resource, was also formed from living matter. Most coal is derived from trees and ferns that grew some 250–300 million years ago during the carboniferous period. This time window was much shorter than that of petroleum formation, which spanned more than 400 million years. Aerobic bacteria generally decompose most plant matter into CO_2, but when a tree fell into a swamp and was buried with limited exposure to oxygen, it could be partially preserved. That phenomenon occurs even today and is evidenced in the formation of peat (very young coal) in bogs.

In the swamps, anaerobic microbial action decomposed most of the tree's cellulose, but the tree's lignin, which was harder to decompose, was saved in the oxygen-depleted environment. Over time, this portion of the tree combined with clays and other minerals and was buried deep under other debris and sediment. Under the influence of moderately high temperatures, the remaining biomass lost most of its oxygen and was transformed into coal. Thus, the older coals, such as anthracites and bituminites, contain less oxygen and have higher heating values than do their younger counterparts such as lignite.

Definitions of Fossil Energy Reserves and Resources

Estimates of the amount of available fossil fuels, and particularly of oil, are often discussed in terms of reserves and resources. Geologists and energy producers ascribe special meaning to these terms,[6] and they often use modifiers such as *proven, probable,* or *possible* to provide further specificity. Sometimes, the reports use probabilistic modifiers such as P95 (indicating a 95% chance of retrieving the stated amount), P50 (indicating a 50% chance), and so on, to denote the probability associated with an estimate of availability. These nuances need to be taken into consideration to avoid confusion. Furthermore, we need to recognize the kind of resource accumulation that is being estimated, whether it is *conventional* or *unconventional, developed* or *undeveloped,* and so on. The degree of certainty with which these estimates are made varies markedly depending on whether the resource is heavily explored and developed, versus being deduced mainly from consideration of geologic features and previous experience with similar formations.

5. Energy Information Agency. *International Energy Outlook,* May 2007. U.S. Department of Energy, Washington, D.C. http://www.eia.doe.gov/oiaf/ieo/index.html. Accessed August 2007.

6. V. E. McKelvey. "Mineral resource estimates and public policy," *American Scientist,* vol. 60, pp. 1–17, 1972.

Finally, the element of economic uncertainty is also an integral part of these definitions: the reported estimates change depending upon economic assumptions. Thus, the quantities of oil and gas estimated within any of these categories can change with changing market conditions or advances in technology that permit a greater fraction to be extracted. Indeed, all estimates of hydrocarbon resources are just that—estimates subject to revision as the price changes and as technology for extraction improves.

In practice, only a fraction of the oil present in a reservoir is extracted. In past years, underground pressure drove oil or gas to the surface in many fields. As the underground pressure was relieved, the oil flow slowed and finally stopped. Some of the remaining oil could then be recovered by pumping. Geological structures and the viscosity of oil determine the economics of pumping and other enhanced recovery measures, such as driving oil upward by injecting water or CO_2. At some point, the cost of extraction technology exceeds the value of oil being recovered and production will be stopped, with some of the oil still underground. In earlier times as much as half of the reservoir's oil was probably left behind when the well (or field) was abandoned because of poor productivity. Nowadays some analysts and producers say that more than 60% recovery can be achieved. Referring back to the most productive oil field, Ghawar, about 50 billion barrels or about 60% of the 80 billion barrel estimate for oil-in-place have already been produced. This 60% figure could be taken either as a good sign—because they are getting well more than 50% of oil-in-place—or as a warning sign, as Simmons has pointed out, that there is perhaps only another 10% or 15% of extractable oil remaining, unless the original estimate of 80 billion barrels is wrong!

Exhibit 5.1, which is adapted from a report by the National Petroleum Council,[7] displays the different categories of resources based on geologic and technoeconomic uncertainties. It divides our resource base into five areas. "Proved reserves" are resources believed to be recoverable using currently known technology at current prices. While the recent rise in oil price has resulted in an expansion of the proved reserves estimates, it has not changed the amount of oil in the ground. The "probable" and "possible" areas along the top of the diagram represent amounts of materials that might be extracted if new recovery technologies are developed, or if the old technologies become economically viable with rising prices. Depending on how future prices of fossil resources rise or extraction costs are lowered, or as drilling shows more or less than the originally projected amounts to actually be present, specific accumulations currently listed as "subeconomic resources" could be reassigned to the other resource categories. Finally, there is presumed to be an unknown quantity of resource that has been neither found nor identified.

7. Committee for Global Oil and Gas, Lee R. Raymond, Chair. *Facing the Hard Truths about Energy,* National Petroleum Council, Washington, D.C., 2007.

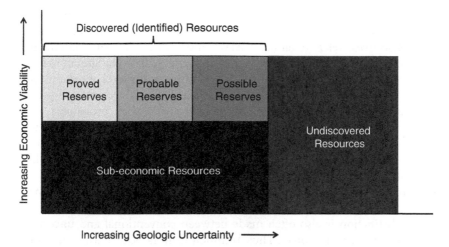

Exhibit 5.1. A Simplified Description of the Prospects for Recovering a Natural Resource from the Ground

Exhibit 5.1 is a simplified version of the McKelvey diagram used in petroleum geochemistry textbooks. In that diagram, reserves, which are the accumulations believed recoverable with existing technology and current and assumed future cost ranges, are nested within boundaries of uncertainties regarding costs of recovery and degree of assurance or certainty of knowledge. Given the floating nature of these boundaries, it is not surprising that considerable confusion and misunderstanding, in addition to the unavoidable uncertainty on probabilities, surround estimates of our fossil resources.[8]

"Proved reserves" refers to accumulated quantities of these inherited energy resources that can be recovered *economically* with *current technology*. As Peter McCabe, a world-renowned geologist, pointed out, this definition may be acceptable to geologists but is puzzling to nongeologists.[9] To begin with, the term does not specify how long the current economic and technological conditions may last. Further, expansion of the reserve base does not necessarily imply the discovery of new accumulations, because expansion may also result from the growth of existing fields when new pockets of accumulation within those fields are located or become accessible through newer technologies, such as directional drilling. With the advent of steerable rotary drill

---

8. Shell Oil Company has posted a skit about reserves and resources on its Web site. While humorously presented, the skit drives home the serious message that these resources are ultimately limited and nonrenewable, given the eons required for their formation. See http://www.shell.us/home/content/usa/responsible_energy/education/energize_your_future/ignite_learning/ignite_resources_reserves.html (accessed September 2009).

9. Peter J. McCabe. "Energy resources; cornucopia or empty barrel?" *American Association of Petroleum Geologists Bulletin,* vol. 82, pp. 2110–2134, 1998.

heads, it is now possible to change the direction of drilling, snake the hole to follow the oil deposits, and recover a much larger fraction of the oil-in-place.

"Resources" refers to all the reserves and additional accumulations that may become technically recoverable in the future, with no judgment made as to whether recovery will be economic. Depending on the level of certainty, the estimates for reserves and resources are further subdivided into (1) proven or assured (95% probability), (2) probable or mean (50% probability), and (3) possible (5% probability). The high-probability resources (95%) represent the lower quantity limit of probable recovery, whereas the possible resources represent the high limit. Another term often found in discussions of these estimates is the "ultimate recoverable resource," which includes the proven resource under consideration, *plus* all the quantities that have already been produced.

A distinction is also often made between conventional and unconventional reserves and resources. Thus, bitumen from tar sands, which is similar to petroleum in its chemical character but whose extraction requires quite different technology, is an example of an unconventional resource. For natural gas, the unconventional resources are tight sands gas, coal-bed methane, and gas hydrates.

As mentioned earlier, it is important to remember that neither resource nor reserve estimates are fixed quantities. Resources and reserves increase as we discover more accumulations and as improved extraction technology becomes available. The current as well as anticipated future market prices of oil also have an influence in determining whether a certain accumulation is part of a reserve or a resource; a resource that was not economical to extract at $25/bbl may well become an extractable reserve at $45/bbl or at $100/bbl. If the purpose of extracting an oil resource is principally to gain energy, then apart from the monetary value, we must also consider its energy value. Once the energy required to extract a barrel of oil exceeds the amount of energy we can derive from it, there will no longer be an incentive to extract it regardless of price.

Superior three-dimensional seismic imaging (made possible by advances in computing) and the development of directional drilling are examples of technical advances that markedly increased global petroleum reserves. Many previously inaccessible pockets within existing oil fields became economically accessible with these technologies. A study by the U.S. Department of Energy states, "Since 1990, the vast majority of reserve additions in the United States—89 percent of oil reserve additions and 92 percent of gas reserve additions—have come from finding new reserves in old fields."[10] These increases, coupled with low oil prices, led some to the general feeling around the turn of the millennium that we were awash in oil.[11]

10. U.S. Energy Information Agency. *Environmental Benefits of Advanced Oil and Gas Exploration and Production Technology,* 1999, p. 16.

11. See, for example, "Drowning in oil," *The Economist,* March 1999.

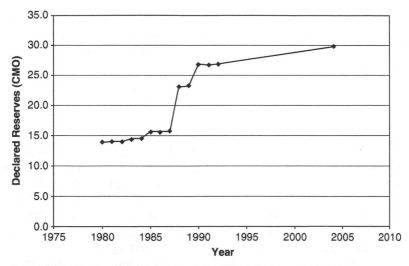

Exhibit 5.2. Declared Oil Reserves of the OPEC Nations, 1980–2005

Subjective factors also contribute to assessments of resource and reserve bases. Analysts making these assessments have to rely on individual company and government reports about the amounts of their holdings. As McCabe has pointed out, a field manager often has the incentive to underreport reserves and production capabilities, because it is always better to face your boss when you deliver more than you promised rather than the other way around. Incentives exist for overestimating reserves, too. As an example, when OPEC (Organization of the Petroleum Exporting Countries) introduced a production limitation quota for its member nations in 1985, it based the quota on the reported reserves of those countries. Member countries thus had a strong motivation to report larger reserves. Exhibit 5.2 displays the large jump in reported OPEC oil reserves between 1985 and 1990—from 15.6 CMO to 23.2 CMO, a whopping 72% increase that occurred essentially overnight, with the discovery of essentially no new fields! Companies may also have motives for overreporting to bolster their stock value, even if they do not act on such motive. In early 2004, Shell Oil Company submitted a revised estimate of its oil reserves to the U.S. Securities and Exchange Commission following an internal audit done to conform to the commission's standards. The revision decreased the estimate of the company's reserves by 18%. Almost immediately the market value of Shell stock dropped 10%, and shortly thereafter some of its senior executives lost their jobs.

As this discussion illustrates, reserves and resource estimates are not hard numbers. They change with circumstances, sometimes by substantial amounts. Accordingly, we must treat these estimates as just that, *estimates*. That does not mean that we should ignore them. Rather, we need to acknowledge the limits of our knowledge and use that information to guide us, just as we do with weather forecasts. A forecast that calls for a 50% chance of showers tomorrow may express uncertainty, but it is much more informative and useful than one that says, "it will either rain or not rain tomorrow"—a forecast that can be made with absolute certainty yet has no utility.

Our Global Inheritance of Fossil Fuels

BP (formerly British Petroleum) has been meticulously compiling data on production and consumption of various fuels by different countries for decades, and its reports are considered very authoritative. It also publishes estimates of global assured or proved reserves of different energy sources. In the following discussion, and in the rest of this book, we have relied on *The BP Statistical Review of Energy* (2007) for much of the data.[12] BP tends to be somewhat more conservative than many other analysts when including accumulations in the proved reserve category. Resources that are inferred but not demonstrated are excluded from the reserves.

Because BP does not list the resource base, we have relied on other sources for those estimates. The U.S. Geological Survey (USGS) published estimates of global occurrences of petroleum and natural gas resources in 1996 and in 2000, and we have used those data here.[13] That compilation includes prospects for offshore oil and gas for all countries except the United States. For the estimates of U.S. offshore oil, we turned to the U.S. Minerals Management Service (USMMS).[14] In contrast to BP, the USGS has been criticized by others who are knowledgeable in the field for being too optimistic in its assessments. To paraphrase Kenneth S. Deffeyes, professor emeritus of petroleum geology at Princeton University and one of the critics, optimists are found in oil companies and in government agencies, whereas pessimists are found among academics and retired petroleum geologists. (The latter two categories are not mutually exclusive.)

For the estimates of our coal resource base, we have relied on a review by Hans-Holger Rogner that was compiled under the aegis of the World Energy

12. *BP Statistical Review of World Energy*, 2007. http://www.bp.com/liveassets/ bp_internet/globalbp/globalbp_uk_english/reports_and_publications/statistical_energy_ review_2007/STAGING/local_assets/downloads/pdf/statistical_review_of_world_energy_ full_report_2007.pdf. This review reports data for 2006.

13. U.S. Geological Survey. *Digital Data Series DDS-30 Release*, vol. 2, U.S. Department of the Interior, Washington, D.C., 1996; USGS World Energy Assessment Team. *U.S. Geological Survey World Petroleum Assessment 2000*, U.S. Department of the Interior, Washington, D.C., 2000.

14. U.S. Minerals Management Service. *An Assessment of the Undiscovered Hydrocarbon Potential of the Nation's Outer Continental Shelf*, OCS Report MMS 96–0034, U.S. Department of the Interior, Washington, D.C., 1997.

Council.[15] Because the different reports do not use the same definitions of reserves and resources and geographic groupings, and because the reports were prepared at somewhat different times, we have had to exercise some judgment in pulling together a consistent set of data for this book.

For each of the fossil resources—petroleum, natural gas, and coal—we first present information on their reserves, followed by information on resources by probability, and finally information on what might be considered as unconventional resources. After discussing the entire resource base of a fossil fuel, we address how long that particular fossil resource can be expected to last under various scenarios.

In contrast to uncertainties regarding the size of our fossil inheritance, we know rather precisely how much we are spending now. We also know that future spending rates are almost certain to rise for an extended period (see chapters 3 and 4). We currently use about 2.4 CMO per year of our fossil inheritance. We also know that shortages of fossil fuels, particularly oil and gas, would result in increased efforts to produce these fuels, and therefore the estimated quantities of conventional fuels available could be augmented by exploitation of unconventional fossil fuels. The ultimate quantity could perhaps be two times or even five times the current estimates. Even so, these *fossil* resources have finite limits and will therefore be exhausted should we continue on a path of unmitigated exponential growth in consuming them. Therefore, it is imperative to begin developing alternate sources of energy now. Other than large hydroelectric power and wood, our income sources contribute a miniscule portion of global energy at present, and they will require a long development time before becoming both cost competitive and scalable to the level of a CMO/yr, if at all.

Global Petroleum Resource

Conventional Petroleum Reserves

Exhibit 5.3 indicates estimated conventional petroleum reserves in the various regions. The term "petroleum" refers to two groups of fossil fuel liquids: (1) oil, which is composed of complex and higher-molecular-weight organic compounds containing mainly hydrogen and carbon atoms, and (2) simpler, lower-molecular-weight liquid organic compounds (natural gas liquids, NGLs) associated with natural gas production, which generally comprise less than 5% of crude oil production. The data in exhibit 5.3 refer to the combined total for the two categories. The most common natural gas liquids in general commerce are the liquefied petroleum gases (LPGs) propane and butane.

15. *World Energy Resources, 2000,* World Energy Council, London, 2001; H. H. Rogner. "An assessment of world hydrocarbon resources," *Annual Reviews of Energy and the Environment,* vol. 22, pp. 217–262, 1997.

Exhibit 5.3.
Proved Global Petroleum Reserves, 2006

Region	CMO	Percentage of World Total
North America[a]	2.3	5.0%
Central/South America	4.0	8.6%
Europe	0.6	1.2%
Russian Group	4.9	10.7%
Asia/Pacific	2.3	4.9%
Middle East/North Africa	30.6	66.3%
Sub-Saharan Africa	1.5	3.4%
World Total[b]	46.1	100.0%

[a] The U.S. portion is 1.4 CMO, or 3.5% of the world total. Geographically, Greenland is a part of the North America region; politically, it is a part of Denmark. Its inclusion in either of the regions has little effect on the totals presented here.

[b] Numbers may not add up because of individual rounding.

As of 2006, BP estimated global petroleum reserves to be 46 CMO (i.e., about 1.2 trillion barrels). In 2000, the USGS published an assessment that covered the world petroleum fields through 1996 and placed the global reserves of petroleum at 36.6 CMO. In the intervening 10 years, the world had *consumed* approximately 10 CMO, yet the reserves had *grown* by about 9.4 CMO. No major oil fields were discovered during this period, and the growth was mostly due to reserve growth that resulted from reassessment of fields under the changed economic and technological circumstances. It is also noteworthy that the recent larger estimate is from BP, the organization that uses a more conservative definition for oil reserves. In the face of this bias toward low estimation, the growth in reserves is even more spectacular: almost 20 CMO was added to the "proved reserves" as a result of the increased ability to access oil in existing or known oil fields.

We see from exhibit 5.3 that the Middle East/North Africa region contains two-thirds of the world's proved oil reserves. The region with the next largest reserves is the Russian group. Russia and Venezuela added together control about 20% of the reserves. On the other hand, the regions where oil is consumed most heavily are currently North America, Europe, and Asia/Pacific. The North America and Europe regions taken together consume almost one-half (47%) of current petroleum production but have only 8% of the world's proved oil reserves. The Asia/Pacific region, which has less than 5% of global petroleum reserves, is expected to become a much larger consumer of oil, given its already large population and projected economic growth. This likely increase is especially the case for China and India, which together account for more than a third of the world's population and have rapidly expanding economies with rapid increases in per capita wealth.

Since petroleum provides essentially all the energy used in transportation, it is worth considering that while in the United States there are approximately 450 cars and light trucks for every 1,000 persons, the corresponding number in China is only 9 vehicles per 1,000 people. Thus, even a growth to 18 cars per 1,000 could translate to a near doubling of China's gasoline consumption, and a likely doubling of the total petroleum consumption from 7.5 million barrels per day (Mbpd) to 15 Mbpd. An increase of 7.5 Mbpd in the face of no new discoveries and aging supergiant fields like Ghawar paint a dire picture. Therefore, it is not surprising that China is vigorously pursuing alliances to gain access to petroleum from any country that has some to spare, even in cases such as Sudan, where the government is unstable and violent.

The geopolitical consequences of the distribution of reserves are manifold. Countries in general try to leverage whatever endowment they may have for economic and political advantage. That is the usual way by which governments secure a better living environment for their citizens. It is this way for oil-rich nations, too, with one potential difference: not all countries in the Middle East/North Africa region enjoy widespread domestic support. Most are currently under autocratic (or "quasi-authoritarian") regimes. Political instability in the region is a major impetus for other countries to seek energy independence. As Thomas Friedman, author and columnist for the New York Times, enunciated in his "First Law of Petropolitics,"[16] at higher crude oil prices these regimes are bringing in all the revenue they need from exports and do not have to tax their people. Accordingly, they have even less incentive to represent the will of the people and institute democratic reforms. Indeed, they can—and have—used this advantage to maintain their regimes.

In attempts to ensure their oil supplies, oil-importing countries often try to seek favor with current regimes, but in doing so they may further exacerbate the very political instability that prompted them to curry favor in the first place. This vicious cycle has no easy resolution, and for that reason banking on the large oil reserve of this region, particularly over a long period, is not a good policy to pursue.

Additional Conventional Petroleum

The USGS and USMMS have estimated the amount of petroleum that might be extracted from yet-to-be-discovered deposits, and how much more might be extracted (i.e., reserve growth) in addition to current estimates of the potential of known oil fields. As mentioned above, the USGS data cover estimates of onshore oil for all countries, and offshore oil fields for all countries except the United States. We took the data for offshore U.S. oil fields from

16. T. Friedman. *Foreign Policy,* Carnegie Endowment for International Peace, Washington, D.C., 2006. http://www.foreignpolicy.com/story/cms.php?story_id=3426 Accessed November 2007.

the USMMS assessments. Exhibit 5.4 summarizes these estimates, which are considered further below.

In estimating the reserve growth around the world, the USGS relied primarily on the U.S. experience, namely, determining the increase in field production observed over decades, gauging how much those increases exceeded the original estimates of recovery, and applying this factor to fields elsewhere. At least partially, that expansion has been due to improvements in extraction technology. The mean global estimate of an additional 35 CMO from reserve growth is approximately three-quarters of BP's 2006 reserve estimate of 46 CMO.

Expansion of conventional oil reserves will also be expected to occur in currently petroleum reserve-rich nations. The Middle East/North Africa region accounts for about one-third of resource additions at both the 95% and 50% estimates. The Russian group and North America and Central/South America regions together account for almost one-half of the total mean estimate of potential new discoveries. Despite differences in the details of these assumptions, the Middle East/North Africa region and Russian group will clearly dominate future petroleum production. However, as noted, both these regions face potential political uncertainties, and global and regional planners must consider what their dominance of overall petroleum resources could mean in future years.

Unconventional Oil Resources

In addition to conventional petroleum recovered with conventional techniques, enhanced oil recovery techniques add to our fossil fuel base. Applying these techniques allows the extraction of additional oil from known deposits

Exhibit 5.4.
Additional Conventional Petroleum Resources (CMO)

Region	Probability of Recovery		
	Assured (95%)	Mean (50%)	Speculative (5%)
North America	2.9	5.7	9.8
Central/South America	0.79	4.7	16
Europe	0.26	1.3	4.5
Russian Group	1.5	6.3	19
Asia/Pacific	0.40	1.9	5.1
Middle East/North Africa	3.0	12	39
Sub-Saharan Africa	1.0	3.1	5.4
World Total	10	35	99

(i.e., growth of reserves). Moreover, new approaches can be used to extract oil from unconventional sources (e.g., from heavy oil and bitumen in sands, or from kerogen in oil shales). The growth in oil reserves forecast in the USGS's 2000 study has primarily resulted from water flooding or from hot water or steam injection. In flooding, pressurized water is injected into an oil reservoir to increase its pressure and thus force more oil to the surface. Hot water or steam injection lowers oil viscosity and thus increases the flow rate of oil to the surface. In a few instances, CO_2 has been injected to increase reservoir pressure and to reduce oil viscosity and thus increase flow. To recover still more oil, surfactants may be injected into reservoirs to increase oil flow through the pores of the formation, thus increasing yield.

The heavy petroleum deposits in Orinoco, Venezuela, and Alberta, Canada, are two unconventional resources that are noteworthy. The heavy oils of the Orinoco oil field are sufficiently heated by geothermal sources to flow to the surface without externally applied pressure. At the surface they cool and become very viscous. In order to pipe them to a seaport for shipping, or to other points of use, they must be mixed with hot water. Although the resulting oil emulsion is sometimes burned in electric power stations as a substitute for coal, overall demand for such a product is small. Heating is also used to extract bitumen (oil) from tar sands in the Athabasca region of Alberta. Depending on the site, these deposits either are mined first and then heated to make lighter hydrocarbons (referred to as retorting) or, if the deposits are near the surface (400–600 ft below ground), are heated *in situ* underground and then collected. In either case, the bitumen so released is processed onsite to recover usable oil. The technology for extracting bitumen from Athabasca tar sands is now fully commercialized. Current operations are producing more than a million barrels a day of synthetic crude oil, which by all accounts is a massive operation. Very few oil fields produce more than a million barrels of oil a day.

Shale is also treated to produce usable oil. Heating mined shale produces oil from its precursor, kerogen. The produced oil can then be refined into products by conventional means. China and Estonia have long produced small amounts of oil from shale. Such retorting of mined shale to recover oil results in a volumetric expansion of the shale, by roughly 30%, and this expansion creates a serious disposal problem. The remaining mineral volume is too large to put back into the mines and as yet has found limited applications. If left on the surface, toxic minerals can leach out of this mineral refuse and contaminate water supplies. The United States has very large oil shale deposits in Colorado and Kentucky, but they are not being mined because of the high costs of producing oil from them and because of potential environmental problems.[17]

17. For example, the Colorado plateau has relatively large oil shale deposits but is also the source of the Colorado River, which could become contaminated by shale processing.

Alternatively, kerogen in shale can be heated *in situ*, underground, using a cavern dug below the shale deposit. Burning some of the kerogen in the cavern releases heat, which in turn releases oil from the remaining kerogen. The unburned liquid remainder is collected at the bottom of the excavation and pumped to the surface. To enhance this method, the U.S. Department of Energy sponsored a novel *in situ* technique that heats the shale electrically and collects the oil given off using pipes emplaced in the body of the deposit. However, a key finding of the project was that electrical heating must continue for several years before substantial amounts of oil could be produced. The study also raised the possibility of the oil so produced migrating into underground water tables and contaminating them.

Recently, Shell Oil announced successful runs of an *in situ* conversion process at the Mahogany Ridge demonstration project in the Green River Valley region of Colorado. As in the aforementioned DOE project, they provided the energy for converting kerogen to oil by electrically heating the shale to 700°F (~350°C). Furthermore, to prevent the liberated oil from dispersing to surrounding areas, they froze the groundwater and created an impervious wall of ice surrounding the region where the conversion process was conducted. About 1,400 barrels of light oil produced in the test; most of the heavy oil components, remained underground. The process allowed Shell to recover about 50% of the oil in the shale at about a 3:1 ratio of energy return on energy invested. The company claims that by this process it will be able to produce high-quality oil at $35 a barrel.

If the Shell process proves commercially viable, the energy structure of the world could shift rapidly. The largest shale deposits are in the Rocky Mountains of the United States, and they are estimated to contain between 20 and 40 CMO. That is two to four times as much as Saudi Arabia has in conventional oil. Even at 50% recovery, the resource potential is huge. The chief barriers to this extraction process are twofold: (a) limitation of water and power resources, and (b) overall economics. There is little water in this arid region of the Rocky Mountains, and residents are apprehensive about the prospect of losing the water they need for drinking and agriculture. Also, the electric grid in the region would be unable to support such a large project, and new power plants would need to be built that would likely increase emissions. As for the economics, Shell's claim notwithstanding, most estimates have placed the recovery to be viable at more than $80 per barrel of crude oil. To overcome the hurdle of initial investment, that price would mean the process would not be viable unless the price of oil is sustained at more than $90 a barrel for about 5–7 years.

Coal can be construed as another unconventional resource of oil, because it can be converted into liquid fuels and supplant oil. Technologies for converting coal to liquids (CTL) have been developed and practiced in several countries, notably during periods of war and economic isolation. Two different approaches have been developed: (1) direct liquefaction, a process in which coal is broken down and liquefied by heating it in a solvent under high hydrogen pressures to temperatures between 700°F and 800°F (375–425°C);

and (2) indirect liquefaction, in which coal is first gasified into a mixture of CO and hydrogen (synthesis gas or syngas), followed by a catalytic process that converts the synthesis gas to the carbon chains of liquid hydrocarbons. Germany and Japan used direct liquefaction to produce liquid fuels during World War II, and indirect liquefaction was developed for commercialization by South Africa.

During the 1970s and 1980s, in the wake of the oil shocks, there was considerable research and development effort expended on improving the economics of the liquefaction approaches. According to a retrospective study on direct liquefaction, these efforts steadily increased the yield of the desired liquid products and decreased the required selling price of crude oil, from $65 per barrel to around $35 per barrel (inflation-adjusted 2000 U.S. dollars). Neither approach is currently competitive in an open market, and the reason Sasol in South Africa is able to operate the indirect liquefaction route commercially is that most of their equipment and construction costs have been fully amortized. Recent sharp increases in construction costs have resulted in estimates for the required selling price of crude oil from a new CTL facility in the range of $75 per barrel. As all the processes for converting coal into oil require energy inputs that are often provided by burning coal, and as coal is lower in hydrogen content than oil, use of CTL will result in an increase in CO_2 emissions over the use of conventional oil by as much as 100%.

Despite the resource limitations and foreboding economic prospects, the promise of vast resources of shale and coal has, once again, attracted many investors and developers. Only time will tell if this time around the promise turns into reality, or will it again end up with bankrupt companies and lost jobs? Too many details about these processes are still unknown to allow us to make a substantive assessment.

What is the extent and availability of these unconventional oils? Recent reports indicate that there is an even larger degree of uncertainty in estimating the reserve and resource base of these unconventional sources (see exhibit 5.5) than those associated with conventional oil. The total reserves of unconventional oil have been estimated to be between 13 and 83 CMO, most of it occurring in the heavy oils. The global resource base of these unconventional sources is estimated to be between 189 and 446 CMO, most of this in shale oils. For comparison, the reserve base of conventional oil is 46 CMO, and the resource base is estimated to be as high as 94 CMO. If we could exploit all these unconventional sources and turn them into oil, our overall resource base would increase many times over.

How Long before We Deplete Our Oil?

Beginning with the 2006 production rate of 1.06 CMO/yr and assuming a continued growth rate of about 1.5%/yr, which was the average growth rate in oil consumption over the 1996–2006 period, the estimated time to exhaust conventional petroleum reserves would be 31 years. At the average growth rate

Exhibit 5.5.
Estimates of Unconventional Oil Reserves and Resources (CMO, rounded)

Oil Source	Reserves		Resources	
	Low	High	Low	High
Heavy Oil	9.1	49	50	96
Tar Sands	0.3	1.7	19	110
Shale Oil	3.6	32	120	240
Total	13	83	189	446

of consumption during the last 40 years (for which we have detailed records, showing 2.4%/yr growth between 1965 and 2005), current conventional petroleum reserves would be exhausted sooner, by 2033. Even if we assume, as the petroleum industry has often done in its public pronouncements, that the rate of petroleum use will remain unchanged from its current level (1 CMO/yr), 46 CMO would last for only 46 years, or to 2052, which is essentially within the time horizon of our study. We conclude that the conventional oil *reserves,* which have been *already identified,* will in all likelihood be largely gone by 2050, which is not a long period in human history. Of course, between now and 2050, some of the petroleum deposits currently classified as a *resource* may become reclassified as reserves, which introduces uncertainty in the estimate of time to exhaustion of *reserves.*

The resource base is considerably larger, and developing unconventional sources can further augment the reserves base. However, there is considerable uncertainty about the true extent of our oil inheritance. In exhibit 5.6, we present the size of global reserves and resources as a pyramid *a la* Peter McCabe. The pyramid has been roughly scaled so that the volumes correspond to the estimates of various classes of the total accumulation. The estimate for unconventional resources has been underrepresented in this figure so as not to overwhelm the other classes . The proved reserves occupy the apex, and estimates of conventional and resource bases at different levels of confidence are shown as layers in the expanding base. Unconventional resources occupy the bottom two layers. We may have only a relatively small amount left (46 CMO of proved reserves), or we may have enough (140 CMO in the conventional resource base and 446 CMO of the unconventional resource base) to last us for centuries. What are the practically recoverable quantities, and how quickly should they be exploited? Should we assume in planning for our energy future that only known reserves will be available, or should we be extremely optimistic and assume that all expected reserve growth and the speculative resources (5% probability) will make much larger quantities ultimately available? How long will it take to deplete the total assumed petroleum resource?

The answers depend in large measure on assumptions about the rates of petroleum use and the available quantities. In exhibit 5.7 we give the calculated

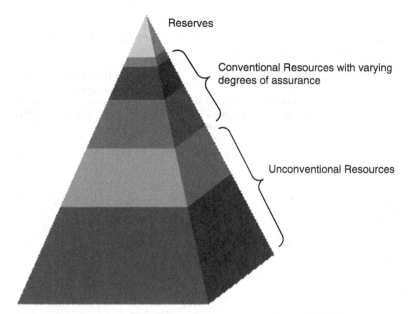

Reserves

Conventional Resources with varying degrees of assurance

Unconventional Resources

Exhibit 5.6. Representing Our Oil Endowment as a Pyramid. Volumes Approximately Correspond to Relative Amounts of Oil in Different Classifications

time for exhausting the reserve or resource under different scenarios based on how deeply we are able to access the conventional resource base. If we assume that the current 1 CMO/yr rate of consumption will continue indefinitely, then our time scale ranges from 46 years for reserves only, to 140 years if we are able to fully recover speculative resources. If we also assume that the "assured" resources (95% probability) will be added to existing reserves, we arrive at the total of 56 CMO, and 56 years to depletion.

However, petroleum use has not been constant over the years. Between 1995 and 2005, it increased at an annual rate of 1.5%/yr. Hence, we also calculated the years to depletion under different scenarios of growth in oil use, by bracketing the recent growth rate with somewhat higher and lower values. The estimates of time to exhaustion of reserves plus resources under these scenarios are significantly reduced, ranging from 31 to 84 years. Beyond the conventional resources, there are vast amounts of unconventional resources. We have presented the range of their estimates, but there are large uncertainties surrounding what fraction of the unconventional resources would be recoverable, so we have not included them in our calculation.

Although the estimates for time to depletion of even the conventional petroleum *resource* base carry us beyond the 50-year time horizon of this book, future planners should not ignore such estimates. Global reserves, plus reasonably assured resources, amount to 56 CMO. That amount results in a calculated time to exhaustion of sometime between 2037 and 2062. If we consider

Exhibit 5.7.
Estimated Years to Depletion of World Petroleum Resources and Reserves
(rounded to two significant figures)

	Quantity Available (CMO)	Current Use (1 CMO/yr)	Years to Depletion, Based on Percent Increase in Use per Year		
			1.0%/yr	1.5%/yr	2.0%/yr
Reserves (2006)	46	46	36	33	31
Reserves, plus "Assured" Resources	56	56	42	39	36
Reserves, plus "Mean" Resources	81	81	56	50	46
Reserves, plus "Speculative" Resources	140	140	84	73	64
Unconventional Resources					
Heavy Oil	50–100				
Tar Sands	20–110				
Oil Shale	120–200				

the reserves and *likely* resources (50% probability), the time to exhaustion ranges between 2052 and 2087. These dates may seem far, but those planning for the medium term must keep in mind the slow rate by which changes occur in large industries.

Peak Oil

The foregoing discussion mostly concerns the size of our endowment. As we alluded to earlier in this chapter, the ability of the petroleum industry to meet the increasing demand is a very important consideration. Temporary shortages and surpluses have historically had a large effect on the price of crude oil, which in turn affects the price we pay for fuel at the pump. Cost of energy, and oil specifically, is an important factor in the price of many other commodities. Low energy prices facilitated the globalization of manufacturing and agriculture, and thus many of the goods and foods we buy come from long distances away where they can be produced more economically. High cost of energy affects that dynamic, and in the short term the impact is generally a rise in the price of goods. Thus, while the size of the resource is important, the rate at which the resource can be supplied is also extremely important.

In the 1950s, L. King Hubbert, a petroleum engineer at Shell Oil, analyzed the production of U.S. oil wells and correctly predicted that the oil production in the United States would peak in 1970. Subsequently, others have applied his method to the global production of conventional oil. They estimate a global peak for oil production, at about 10% more than our current level of consumption, to occur between 2005 and 2015. The essential elements in Hubbert's analysis are that (1) there is a finite resource base, and (2) larger and easier-to-exploit portions of this resource base will be found and developed first. Therefore, finding new fields and producing oil from them gets progressively more difficult with time, causing overall production to begin to decline. Both of Hubbert's statements are reasonable, and the analysis leads to a bell-shaped curve for the production rate over a period of time. Such a curve describing production versus time peaks at a point when half of the resource is exhausted.

Critics point out that while the quantity of the resource is "finite," its size remains uncertain. In the past quarter century, the estimate of global oil resources has more than doubled. Furthermore, there are even larger unconventional resources (e.g., oil shale) that are essentially producing the same commodity (oil). Therefore, it is not clear whether Hubbert's type of analysis can be extended to these unconventional resources, for which very few historical data are available. But if we do accept Hubbert's analysis as applicable, then as long as the demand for oil is high and the production rate keeps increasing, even a quadrupling of the total resource base of oil delays the peak by no more than two decades. It follows, then, that in order to substantially postpone the point of world peak oil production and extend supplies for a long time, we will have to reduce the demand growth from its current exponential rise.

The *rate* at which oil can be produced to meet the expanding demand has a critical and immediate impact on the price. Installing requisite infrastructure for producing oil from unconventional oil has barely begun, and we can all agree that the global peak production of *conventional* oil will occur soon (if it has not already done so). The run-up in crude oil prices between 2006 and July 2008 has many causal factors, and various parties have been vilified: "greedy" OPEC, large multinational oil companies, rampant economic growth in India and China, and also the hedge fund investors. While all of these factors have contributed, the fact that there is less than 3% slack between the demand and production capacity of oil has made every user nation nervous, sent them scrambling to secure their supplies, and also invited investors to use oil as a hedge.

Such run-ups, of course, likely cause a slowing down, or "demand destruction," as it is often referred to in the industry, which reduces the tension in oil supply. The consumption of oil in the United States dropped by about 500,000 barrels per day (bpd) in late 2008 at least partially in response to high prices. In an effort to promote the expansion of their economies, countries such as China and India had been subsidizing the price of oil and not recovering the full cost of imports. Yet even these subsidizing countries were finding it harder to maintain that low internal price of oil, which led to increasing the price the public paid at the pump, which in turn decreased the public's demand. For example, this reduced demand was probably a major factor leading to the retreat in oil prices from a high of $147 a barrel in early July to $113 a barrel by mid August of 2008, before settling to around $40 by November 2008. The point of note here is that under a tight supply situation, a change in demand of even 100,000 bpd (ca. 1% of the global production) can have a substantial effect on the price of oil.

Facing Realities

For various economic, technical, environmental, and political reasons, not all conventional sources of oil in the world have been tapped. In the United States, great debates have arisen over the future of the Arctic National Wildlife Refuge (ANWR) in Alaska, which is rich both in endangered species and in oil, and over offshore drilling, which many worry would damage the coasts and the coastal tourism industry. Proponents of drilling in these places argue that production of this oil is needed as a significant addition to the national oil reserves and would bring gasoline prices down. Those opposed to producing oil in ANWR and offshore claim that such drilling would have little if any effect on gasoline prices and that any oil produced would be a drop in the bucket on the global or national scale.

Both sides have reasonable support for their arguments. However, it is important not to get so absorbed in details or so engrossed by the general arguments that we lose perspective on key specifics of the matter—it is a balancing act. The predictions for the amount of oil in ANWR were made mainly by government agencies between 1980 and 1998. The estimates for oil-in-place increased over the 18-year span from about 5 billion barrels to more than 20 billion barrels (less than 0.2 CMO to 0.8 CMO). An assessment by the American Association of Petroleum Geologists, an independent non-profit organization of experts, basically concurred with the higher assessment. However, when we consider the estimates of amounts judged to be technically recoverable, the numbers are between 0.2 and 0.4 CMO. Compared with the current U.S. consumption, this amount would correspond to between a 10- and 20-month supply, a trivial amount in the larger picture.

On the other hand, we should recognize that total oil produced from the Alaska North Slope also corresponds to a little more than 0.5 CMO. Commercial production in Prudhoe Bay began in earnest in 1977, after the Alaska pipeline was completed, and a peak production of 2 Mbpd occurred 10 years later in 1987. The estimated ANWR reserves are roughly comparable in size, so if we are to dismiss the ANWR potential as insignificant, we should also admit that the production from the Alaska North Slope region has been insignificant! Since most consider the Alaska North Slope oil to have made a significant contribution to the local and the U.S. economy, the ANWR oil, though small in a global perspective, would likewise be a substantial contributor to the economy.

The real question we need to answer is whether the global impact of production from ANWR can be significant. We must remember that any oil that we begin to drill for today will likely not be available in large amounts for at least 10 years. It could easily take that long for the process of site selection and permitting in an environmentally sensitive area, let alone the physical construction. The EIA estimates that ANWR would produce on average approximately 82,000 bpd—for about 18 years, and that productivity has the potential to decrease world price of a barrel of oil by about a dollar a barrel, depending on how tight the supply happens to be at that time.

We also need to consider the environmental impact and the impact on the people who would be most affected by this decision. Would drilling in ANWR change its pristine nature, either temporarily or forever? How much do we value that? The answer to that question has to be a part of the equation. Drilling in ANWR will disrupt caribou migration to their calving grounds and thus affect their population. It would therefore jeopardize the habitat of caribou and the Native American tribes of northern Alaska that depend on caribou as a primary food source. On the other hand, other tribes who do not depend on caribou may have a different response. They may find that the development of ANWR affords employment opportunities for several decades for their people and therefore favor drilling. Some analysts have claimed that much of the oil from what is expected to be ANWR's most productive area can be accessed by horizontal drilling from the wellhead regions of an adjacent field that is already in production. If this is true, then the environmental risks to at least that particular area would be greatly reduced. Alaskans also receive a percentage of the state's oil profits each year. It should not come as a surprise that there will not be a monolithic response to the question of drilling in ANWR.

The situation is similar when it comes to offshore drilling, which by the way has been in practice for nearly 40 years in the Gulf of Mexico, off the coast of California, in the North Sea, and in the Persian Gulf. With one exception, these operations have not led to any major environmental disasters. The exception is the 1969 oil spill in the Dos Cuadras offshore field near Santa Barbara, California, which resulted in leaking an estimated three million gallons of crude oil and contaminating the ocean, blackening the beaches, and killing thousands of birds, seals and other wildlife. It took more than three weeks to contain the disaster. Since 1969, the environmental record of offshore oil production has been remarkably clean, and 40 years of a good record is notable when considering our choices.

In the United States, the current debate is about opening areas in Northern California and the Northwest, in Florida, and on the Atlantic Seaboard to drilling. Off the coast of Texas and other seaboard states, offshore drilling is permitted. The EIA estimates that we could obtain 2.2 million bpd in 2030 with the current offshore drilling already in place, and that drilling in currently restricted areas would increase that to 2.4 million bpd. Is this difference of 200,000 bpd significant? We noted earlier in this chapter the significant impact that a change of 100,000 bpd of production or consumption can have on the oil prices. Also, as with Alaska, the response of people to increasing oil supply strongly depends on what else they value. Citizens along the Florida coast, for example, might be forced to choose between maximizing the output of oil and ensuring that their flourishing tourist industry is not endangered by pollution or the view of rigs.

For national policy makers, the alternatives to drilling in ANWR or in currently restricted offshore areas lie in finding ways to either reduce demand by about 200,000 bpd or produce a comparable amount of oil from alternate sources. We noted the prospect of shale oil earlier. Similarly, CTL technology is an option that has been waiting in the wings for a while and, as we shall see below, the United States has coal resources that could last for centuries. However, for CTL to be economically viable, crude oil prices have to be around $75/bbl. Capital expenses account for the bulk of this $75 price and present a substantial hurdle for any new CTL venture. Should we implement a policy to provide price guarantees for producing oil from coal and shale? Introducing these capabilities would certainly have a more lasting impact on the security of energy supplies than either ANWR or offshore oil. But, they would also almost certainly entail greater CO_2 emissions, unless or until cost effective ways of capturing and sequestering CO_2 are also developed.

In response to the global economic slowdown and plummeting of oil prices in the latter half of 2008, many oil exploration and production projects were abandoned or put on hold. As it takes many years to develop these fields, shutting down these projects should concern us as it inevitably points to a future spike in oil price once the global economic crisis passes.

Natural Gas Resources

Global Natural Gas Reserves

In 2006, BP placed the world's conventional gas reserves at 42 CMO, as shown in exhibit 5.8. The EIA, quoting the *Oil and Gas Journal,* also assumed 42 CMO—up from its earlier estimate of 31 CMO.[18] The earlier Rogner compilation showed gas reserve estimates ranging from 30 to 34 CMO. These differences are not significant for a 50-year planning horizon. At the current (1996–2006) 2.4%/yr rate of increase in gas use, the 42 CMO in gas reserves would be exhausted in 36 years.

The global distribution of gas reserves is also something to give us pause. Two regions, Middle East/North Africa and the Russian Group, have large portions of the world's reserves, 45% and 32%, respectively. As in the case of oil, the large consuming regions—Europe, North America, and Asia/Pacific—are poorly endowed with gas. North America, which is a very large gas consumer, has less than 5% of the world's gas reserves, Asia/Pacific has about 8%, and Europe has about 3%.

18. "Worldwide Look at Reserves and Production," *Oil & Gas Journal,* Vol. 104, no. 47 (2006), pp. 24–25.

Exhibit 5.8.
Proved Global Natural Gas Reserves, 2006

Region	CMO	Percentage of World Total
North America[a]	1.8	4.4
Central/South America	1.6	3.8
Europe	1.4	3.2
Russian Group	13.4	32.1
Asia/Pacific	3.4	8.2
Middle East/North Africa	18.7	44.8
Sub-Saharan Africa	1.5	3.5
World Total	42	100

[a]The U.S. portion is 1.1 CMO, or 3.1% of the world total.

This uneven distribution in natural gas deposits is even more important than that for oil, because transporting gas over long distances in a pipeline requires considerable energy for pumping. Gas can be transported in tankers as compressed natural gas or liquefied natural gas, but these processes entail expenditure of energy that detracts from the value of the gas. Energy losses associated with compressed natural gas are typically 5–8%, and the liquefied natural gas route costs about 15% of the energy contained in the natural gas.

The gas can also be converted into a petroleum-like product. This gas-to-liquids (GTL) conversion process produces a readily transportable product, but the process entails a substantial energy loss. The process entails two steps: (1) the conversion of natural gas into syngas—a mixture of CO and hydrogen—by reaction with water, and (2) the catalytic transformation of the syn gas into an array of liquid hydrocarbons. Both the steps are well-developed technologies, but capital for the process tends to be expensive (tens of billions of dollars for constructing a plant) and practical only at very large scales. The crude oil price at which producing liquids from gas becomes economical depends also on the price of gas. Gas produced at remote sites, far from consumers, is priced much lower than pipeline gas. The former may be procured at $0.50/MBtu when the latter is being sold as high as $8.00/MBtu. If natural gas can be procured at $0.50/MBtu, a GTL operation could be economically competitive if the crude oil price is between $30 and $50/bbl, depending on the financial terms the sponsors of the project are able to secure. In only a few instances is the amount of natural gas produced at a remote location large enough to warrant those expenses. Since the 1990s, only a handful of large GTL facilities have started operating commercially, in Malaysia, Qatar, and Bahrain. Together, they produce on the order of 100,000 bbl/day of oil, which is an impressive amount, but one that corresponds to only 0.0015 CMO/yr, or 0.25% of total natural gas use.

Additional Conventional Natural Gas

Agencies in the United States have estimated that potential additional world production (reserve growth) from known sources of gas would amount to 34 CMO. As displayed in exhibit 5.9, the Middle East/North Africa region and the Russian group are potentially important sources of additional natural gas, albeit less important than they are for petroleum. Middle East/North Africa holds between 2.8 CMO of assured (95% probability) and 17 CMO of speculative (5% probability) undiscovered natural gas. The Russian group has slightly higher additional sources of natural gas, between 2.8 and 21 CMO for the respective probabilities. They hold between a quarter and a third of the world's additional natural resources, respectively. In contrast, Europe holds an estimated potential for additional discoveries between 0.3 CMO (95% probability) to 4.8 CMO (5% probability). The mean value of 2 CMO (50% probability) corresponds to only about 6% of the world's estimated but yet to be discovered conventional gas deposits. The North America region possesses 2.7 CMO of assured growth natural gas resources (which are 150% of its current reserves). The assured reserve growth in North America is essentially the same amount of high-probability gas resources as present in the Middle East/North Africa region and Russian group, but North America has considerably less natural gas in the speculative category (6.9 CMO vs. 17 or 21 CMO, respectively).

Unconventional Natural Gas Resources

Natural gas is most commonly found along with petroleum, but it is also found in separate deposits. For instance, Iran's natural gas reserves are far larger than Saudi Arabia's natural gas reserves, even though Iran's oil reserves are substantially smaller than Saudi Arabia's. In addition, there are accumulations

Exhibit 5.9.
Additional Conventional Natural Gas Resources (CMO)

Region	Probability of Exceeding the Amounts Stated		
	Assured (95%)	Mean (50%)	Speculative (5%)
North America	2.7	4.4	6.9
Central/South America	0.63	3.2	7.1
Europe	0.29	2.0	4.8
Russian Group	2.8	10.5	21
Asia/Pacific	0.91	3.2	6.5
Middle East/North Africa	2.8	8.9	17
Sub-Saharan Africa	0.54	1.5	2.9
World Total	11	34	66

Exhibit 5.10.
Estimates of Unconventional Natural Gas Resources (CMO)

Gas (in place)	Resource
Coal-Bed Methane	60
Fractured Shales	110
Tight Formations	49
Gas Hydrates	5,000

of natural gas associated with coal, shale, and sandstones; these are typically considered as unconventional resources. Another resource that came to light in recent decades is methane gas associated with ice, which is also referred to as gas hydrate.[19] Gas hydrates exist under a very limited range of conditions of temperature and pressure. They can be found under the seafloor or in the permafrost regions. They look like typical ice crystals, but when brought to the surface the crystals release natural gas, which can be lit with a match for a dramatic visual of burning ice.

The potential efficiency of methods for recovering additional natural gas from unconventional sources is uncertain. The estimates of unconventional gas are shown in exhibit 5.10. The estimates shown are for total in-ground amounts of unconventional sources.[20]

Some estimates of natural gas present in the ground in the form of gas hydrates suggest that the amount stored in hydrates might be as large as 1,000 CMO, perhaps even more. While the fraction of gas from hydrates that might be recoverable is unknown at present, even a 10% recovery from deposits this large would solve energy problems for centuries to come. Harvesting them economically has proved very challenging because of the severe conditions under which they are found, and the fact that because the gas is bound as a solid, it does not simply flow to the surface. In recovering natural gas from hydrates, guarding against uncontrolled release is also critical, both to avoid wasting the resource and to protect the environment. Molecule for molecule, natural gas (methane) is about 20 times more potent a greenhouse gas than CO_2. Even though atmospheric processes convert methane, thereby reducing its global warming potential, the ultimate conversion prod-

19. Gas hydrate crystals have one molecule of methane loosely linked to about seven molecules of water. Their formation in under-ocean deposits requires pressures of about 500 psi (or depths of 1,000 ft) and a temperature of 32°F (0°C) or lower. Land-based deposits are likely to require an earth overburden of at least 100 ft and maximum temperatures in the range of +5°F to −14°F (−15°C to −25°C). At a depth of 1,000 ft, the pressure would be great enough that the temperature could be as high as 32°F (0°C) without release of methane. Because earth temperature rises with depth, hydrate formation is limited by depth.

20. For the unconventional natural gas resources, we do not distinguish between reserves and resources, since most unconventional natural gas is currently largely untapped.

uct is CO_2. The extremely large quantities of methane that might get released from gas hydrates, either during production or as a result of general warming, could have a very large impact on the total greenhouse gas concentration in the atmosphere.

Substantial amounts of gas also could be extracted from coal seams, some shales, and "tight" sandstone formations (i.e., those with very small pore size through which gas will move very slowly). The amount of natural gas present in coal seams depends on the type of coal formation, and the gas can be extracted using conventional drilling techniques. Certain shales contain gaseous organic matter that can be recovered by similar drilling techniques. Natural gas could be retrieved by fracturing the formation, usually with explosives, and holding the fractures open by "propants" that prevent small seams in the rock from closing so that a sufficient amount of the contained gas can flow at a reasonable rate toward a collection well.

How Long before We Deplete Our Gas?

The proved reserves of natural gas, plus additional production from known resources, are similar to those for oil. The world's current rate of consumption of natural gas is only about 60% that of petroleum. As a result, natural gas is likely to last longer as a fuel. As shown in exhibit 5.11, the full exploitation of our conventional natural gas resources could, at current rates of use, provide

Exhibit 5.11.
Estimated Years to Depletion of World Natural Gas Resources and Reserves (CMO)

	Quantity Available (CMO)	Current Use (0.6 CMO/yr)	Years to Depletion, Based on Percent Increase in Use per Year		
			2.0%/yr	2.5%/yr	3.0%/yr
Reserves (2006)	42	70	44	40	37
Reserves, plus "Assured" Resources	53	88	51	47	43
Reserves, plus "Mean" Resources	76	127	63	57	52
Reserves, plus "Speculative" Resources	108	180	76	68	62
Unconventional Resources					
Tight Sands	50–100				
Shale	20–110				
Coal Bed	120–200				
Gas Hydrates	1,000?				

gas for 180 years. In that case, one may ask, why worry about natural gas supplies?

The answer matters because the consensus is that natural gas use will rise more rapidly than that of oil, because gas is cleaner and is also more efficient in combined-cycle electric power production or in natural gas vehicles. At the 2.5%/yr growth rate for natural gas, which is similar to the rate of increase of overall energy use over the past century, the calculated maximum time to exhausting gas reserves and assured resources is only 47 years, and there are only 68 years remaining before reserves plus the speculative gas resources are used up. Then we would presumably look to the unconventional natural gas resources. Even if these large amounts of gas resources were to become available, we would exhaust 1,000 CMO in about 150 years at the present rate of expanding consumption (2.5%/yr). This prospect for exhaustion of even a very large fossil fuel resource base shows the inexorable result of continued exponential growth, making it abundantly clear that fossil fuels cannot be a *long-term* (centuries long), large-scale energy source.

Coal Resources

Global Coal Reserves

Coal is not a homogeneous substance—it has many different chemical compositions, impurity contents, and heating values. These variations in coal are significantly greater than those for oil, natural gas, or the low-molecular-weight liquids sometimes extracted from an oil well. Coal grades include anthracite, bituminite, sub-bituminite, and lignite. Peat is a very young coal that is sometimes included with lignite in resource compilations. Each coal grade has a substantial range of energy content. For convenience in energy resource estimations, these grades are often combined into the "hard coal" and "brown coal" categories, whose average energy contents are taken to be 33 MBtu/ton and 18 MBtu/ton, respectively. In contrast, energy content (on a weight basis) for various grades of oil varies by no more than a few percent. Although the energy content of wellhead natural gas may differ by as much as 50% in a few instances, natural gas can be purified to a consistent energy content in a relatively straightforward manner before it enters the transmission system.

Many analysts have estimated coal reserves. Rogner estimated total reserves at about 140 CMO. Exhibit 5.12 shows some of the more recent estimates by BP, which place the total energy content of the coal reserve at 121 CMO, of which about two-thirds is in the more desirable (higher energy content) hard-coal category. Analysis of the BP data shows that, on the basis of energy content, the Asia/Pacific region has more than one-third of all current coal reserves. North America (principally the United States) has about one-quarter. The Russian group has about one-fourth, with all the other regions combined

Exhibit 5.12.
Proved Global Coal Reserves, 2006

Region	Hard Coal (CMO)	Brown Coal (CMO)	Sum (CMO)	Percentage of World Total
North America[a]	20.2	12.1	32.3	26.7
Central/South America	1.3	1.1	2.4	2.0
Europe	3.0	2.1	5.1	4.2
Russian Group	16.6	13.1	29.7	24.6
Asia/Pacific	33.6	9.1	42.7	35.3
Middle East/ North Africa	0.1	0	0.1	<0.1
Sub-Saharan Africa	8.8	0.04	8.8	7.2
World Total	83.6	37.4	121	100

[a]The U.S. portion of all coal is 31 CMO, which is more than 25% of the world total.

having about one-sixth of coal reserves. The Middle East/North Africa region has essentially no coal reserves, and those in Europe are meager.

At 121 CMO, coal reserves are about three times as large as either petroleum or gas reserves. Coal use had been relatively constant for some time, at about 0.6 CMO/yr, but lately it has increased sharply to 0.8 CMO/yr. The 121 CMO of coal reserves could last for more than 150 years at the current consumption rate of 0.8 CMO/yr. The Asia/Pacific and North American regions could alleviate shortfalls in petroleum and gas by converting coal into oil or gas, although doing so would result in a 25–30% loss of useful energy.

The abundance of coal is not necessarily a comfort, given that it entails several drawbacks:

- Coal utilization processes convert the sulfur and nitrogen in coal to gaseous sulfur and nitrogen oxides. The oxides of nitrogen are also formed from the nitrogen in air during combustion of even nitrogen-free fuels. These oxides are acidic products and fall back to the ground as acid rain and thus harm the environment. Their release should be controlled, but too often it is not.
- Coal combustion creates particulate products that contribute to air pollution and, along with the acidic products, are often not controlled despite knowledge of adverse impacts on health.
- Coal use creates more of the greenhouse gas CO_2 than either gas or oil on a per unit of energy provided basis.

Coal also contains toxic materials such as arsenic, lead, mercury, and selenium in various forms, along with radioactive elements thorium and uranium and their decay products. To help protect the environment, the U.S. government currently requires safe storage of coal ash for 30 years, "if feasible." Many countries that burn coal do not have even such optional environmental safeguards in place. Moreover, coal-burning power stations are not the only sources of these impurities: coal-mining residues, which are infrequently controlled, also contaminate the environment. In many countries, coal is burned for household heating and cooking, often in open fireplaces and ovens. This practice presents a significant health hazard to the residents. For all of these reasons, utilization of coal is limited compared to its availability in some countries such as the United States.

With all these environmental drawbacks, one might think that there can be no case made for the use of coal. Actually, it depends on where you are starting from. Switching to coal from firewood is beneficial both for the environment and the health of the individual. Firewood use results in deforestation, loss of ecosystems, and attendant loss of soil quality. Wood fires are also sooty and the particulates contribute to global warming by decreasing the reflectivity of the snowfields. Firewood is generally burned in homes with poor ventilation, and poses a greater health risk to the individuals than they would experience from a coal-fired utility that provides electricity to their homes.

We need cost-effective methods for curbing the emissions so that coal can be deployed more widely. Use of more efficient methods in coal combustion processes, such as supercritical and ultra-supercritical boilers and combined heat and power generation, can ameliorate the negative effects. As we mentioned earlier, a typical coal-fired power plant has a thermal efficiency of only 30%. These advanced boilers can have efficiencies as high as 45% and thus deliver the same amount of energy for a third less coal.

Another way of utilizing coal is through gasification. In that process, part of the coal is burned, using a limited amount of oxygen, and the heat released produces combustible gases such as CO and hydrogen, plus a solid coal char. This char is further gasified with steam to produce more CO and hydrogen. The gases are then burned in a gas turbine to produce power, or they could be used to prepare fuels. The advantages of gasification are several. It can lead to substantially higher overall efficiencies for power production. If the gas turbine is followed by a steam turbine, as in an integrated gasification combined-cycle (IGCC) unit, the overall efficiency of power production can be as high as 60%. The gasification route also leads to significantly lower toxic emissions, and it also offers an easier opportunity to capture CO_2 than in conventional coal-fired power plants. The IGCC technology was developed more than 30 years ago yet is still not practiced for commercial-scale power production, in part because the most energy-efficient way of using coal gasification requires some expensive scrubbing of contaminants from still-hot gases so the gases can be optimally used in high-temperature gas turbines. The high temperatures and corrosive nature of the gases require the use of very expensive corrosion-resistant alloys, but even the special materials have been known to fail, leading to unscheduled shutdowns. These reliability concerns have hindered the adoption of IGCC for power production.

Additional Conventional Coal Resources

For the estimates of coal resource base, we have taken data from Rogner, who as mentioned above had compiled an assessment of global fossil resources in 1997. Rogner divided both the hard coal and brown coal accumulations into five categories (A through E) with successively increasing uncertainty as to their geologic occurrence or technoeconomic viability. His definitions are somewhat different than BP's, and he included in his categories A and B resources that BP would have excluded from "proved reserves." In exhibit 5.13, we report the sum of Rogner's categories A and B as assured reserves; they may be thought of as proved reserves plus 95% probable (assured) resources. We have taken Rogner's categories C and D as indicative of likely resources (50% probability), and category E as speculative resources (5% probability). As before, our numbers in exhibit 5.13 also take into account the differences in energy content of hard and brown coals and give the total resource figures in units of CMO.

The total in the three categories amounts to more than 1,600 CMO, of which about 200 can be considered assured. Of these assured coal resources, about 20% are located in North America, 15% in the Russian group, and about 50% in Asia/Pacific. Europe has about 12% of the assured coal resources, and the two remaining regions contain negligible amounts. If we sum the assured and likely resources, the resource base of coal for the seven regions in declining order is Asia/Pacific (40%), the Russian group (30%), North America

Exhibit 5.13.
Additional Coal Resources (CMO)

Region	Resource Base (Rogner)			
	Assured Reserves (95%)	Mean (50%)	Speculative (5%)	Total
North America	39.6	60	130	230
Central/South America	2.4	2	8	12
Europe	22.7	28	59	109
Russian Group	29.7	149	573	751
Asia/Pacific	91.9	148	241	481
Middle East/ North Africa	0.1	1	3	4
Sub-Saharan Africa	9.6	14	17	40
World Total	196	402	1,030	1,627

(17%), and Europe (8%). Of the three remaining regions, sub-Saharan Africa has about 4%, and the other two regions have negligible amounts.

Although the total quantity in the speculative categories amounts to an additional 1,030 CMO, we do not believe they would change any planner's viewpoint because the very large combined coal assured reserves and likely resource base already amount to 600 CMO. Coal resources are so much larger than those of oil and gas—about 10 times larger in terms of contained energy—that the potential times to exhaustion of the various coal resource bases are at least several centuries. At the current rate of use (0.8 CMO/yr), the 121 CMO of proved coal reserves should last more than 150 years, and assured reserves (196 CMO) for about 250 years. If we assume a continued annual growth rate of 2.0%/yr, the assured reserves would last through the end of the twenty-first century, and the likely reserves can last for additional 50 years.

Unconventional Coal Resources

Still more coal may exist in deposits that are difficult to mine by traditional means or by even today's advanced methods. Those further deposits may be (1) within extremely thin seams separated by thick layers of rock, (2) deeply buried, perhaps at 8,000–10,000 ft or even deeper, or perhaps (3) finely disbursed in rock or soil. It might be possible to recover energy from these difficult-to-access deposits by converting the coal into a low-Btu-content gas—a mixture of hydrogen, CO, and nitrogen—by pumping air down to the deposits, igniting the coal, and using an exit well to recover the resulting gases. Underground coal gasification has been explored previously (largely in Russia) but does not lend itself readily to control, even without considering the escape of carbon oxides.

Infrastructure Requirements for Using Our Fossil Fuel Inheritance

As we consider different sources for our future energy needs and the potential trade-offs, we have to know what else, besides the resource itself, will be needed to develop it and produce a sizable amount of energy from it. The production of energy, no matter what kind of resource is involved, entails major requirements of infrastructure and land. For example, to use oil, we need not only wells for production but also pipelines, tankers, refineries, and other infrastructure. There are environmental impacts associated with every step from production, distribution, refining, and use. The size that we have chosen for making comparisons is a CMO. That is the scale at which any resource will begin to make a significant impact on our overall energy use.

In the following pages, we set forth some estimates of infrastructural requirements and potential environmental impacts, in qualitative terms for environmental effects and in semiquantitative terms for facilities and land requirements. Because much of the required data were unavailable, we have made several assumptions in arriving at these estimates. Therefore, the values presented here should be considered as indicative rather than absolute. Chapters 6 and 7 present similar assessments for the actions necessary to obtain an amount of energy equivalent to 1 CMO/yr each from nuclear and various energy income sources.

Requirements and Potential Impacts of Producing 1 CMO/yr from Oil

The 1 CMO/yr of oil produced in the world annually comes from 850,000 wells. If we simply divide the total amount of oil produced by the number of wells, we get an average productivity of about 130 bbl/day. The productivity of these wells, however, varies over a large range. The United States has approximately 550,000 wells, most of which are "pumpers," producing perhaps only 3 bbl/day. In contrast, certain wells in the Middle East produce more than 30,000 bbl/day or about 10,000 times as much as the pumpers.

If we assume that each well requires an acre of land surrounding it that could not be used for other purposes, then the total areal requirement is 850,000 acres, or about 1,300 square miles.[21] When considering land requirements for production of an *additional* CMO/yr, we will (1) neglect the very small wells and (2) use an average productivity of 300 bpd (which is the geometric mean between pumpers and high-productivity wells). Exhibit 5.14 sets forth these and related requirements for producing 1 CMO/yr from these 300 bpd wells. We would need about 400,000 of such wells requiring about 600 mi^2 of area. The areal requirement is on the order of 1,000 mi^2. It is not small, as in the 100 mi^2 that would be required for nuclear power, nor is it as huge as a million square miles, which, as we shall see, would be needed for producing 1 CMO from energy resources such as wind and biomass.

Pipelines are the most common means of transporting crude oil and refined products.[22] About 90 pipelines, each averaging 800 miles in length and 36 inches in diameter, would be required to transport 1 CMO/yr of crude oil. Ocean tankers transport about one-third of the crude oil used today, traveling about 5,000 miles one way (and returning empty) for about 12 trips per year. These vessels also require special loading and unloading facilities. To transport 1 CMO/yr of oil, a fleet of 280 large-capacity vessels would be needed, each about 350,000 dead weight tons. One tanker costs upward of $10 million, and the fleet would cost several billion dollars. A large, but mostly hidden, cost

---

21. Since the world currently produces 1 CMO/yr of oil, the data for areal and other infrastructural requirements for producing it should be known. However, we could not find that data and therefore resorted to making an estimate.

22. Refined products are also transported in smaller seagoing vessels and river barges.

Exhibit 5.14.
Requirements for Producing 1 CMO/yr from Oil: Selected Operations
(estimates rounded)

Activity	Facility Requirements (Units)	Environmental Impacts	
		Area (mi²)	Other
Extraction			
Exploratory wells	5,000–10,000	8–16	Blowouts, spills, flared gas, toxic drilling materials
Production wells	390,000 (3,900 fields)	600	
Transportation			
Ocean tankers	280	20–40[a]	Transfer spills, accidents, leaks, fires
Crude oil pipelines	90	700	
Product pipelines	1,200	600[b]	
Refining	250	100–150	Fires, accidental releases (mainly of refinery by-products)

[a]Includes 30 export and 50 import docking facilities for tankers carrying 0.3 CMO/yr.

[b]The typical product is a mixture of gasoline, kerosene, and diesel fuel; input and output facilities are included.

incurred by the society is the military investment needed to keep the sea lanes open for transporting oil.

In all of these transportation steps, the energy loss is relatively small compared with the approximately 5–7% loss of the original energy in the oil during refining. About 250 refineries, with a nominal input of 300,000 bbl per day of operation, would be required to refine 1 CMO/yr of crude oil. At currently prevailing prices ($30,000 per daily barrel), each of those refineries would require an investment of about $9 billion dollars. Thus, investments in refineries alone would run into several trillion dollars.

In addition to the direct use of land, there are several other land- and water-related issues. There is the potential for contamination of land through the use of drilling muds and the extrusion of saltwater or brine through wellheads. Accidental releases during normal production, transportation, and refining operations could also result in contamination of land and water.[23] Additionally, the consumption of 1 CMO/yr of oil releases approximately 12 billion tons of CO_2 to the atmosphere.

The total cost of producing a cubic mile of oil is substantial. The component costs vary considerably depending on the productivity of the wells. Petroleum geologists speak of two cost elements in oil production: (1) cost of exploring and developing the field (finding), and (2) cost of producing the oil (lifting). The lifting costs are easier to estimate and range from less than $2 a barrel in parts of Saudi Arabia, to around $6 a barrel for most onshore production, to more than $20 a barrel for offshore operations. According to the EIA, estimates for the cost of finding and developing the fields range from $5 in the Middle East to more than $60 for offshore United States. At these costs the overall investments needed to maintain the productivity of a cubic mile of oil per year are on the order of several hundred billion dollars annually. The International Energy Agency announced in October of 2008 that the cumulative investments through 2030 needed to keep oil productivity projected in their reference scenario (1.6%/yr growth) total $6.3 trillion.

Requirements and Potential Impacts of Producing 1 CMO/yr from Gas

Much of the gas produced today is associated with oil production. Wells producing gas alone are responsible for only about 20% of total gas production, and will likely remain at this level for some time to come. We can use today's infrastructure, which exists for current gas use, to estimate the additional infrastructure required to provide an additional CMO from natural gas. If we assume that there are 1,300 gas-only fields with 100 wells each, producing on average 650,000 ft³/day of gas per well, the annual production would amount to 0.2 CMO of nonassociated gas (i.e., gas that is not associated with oil drilling). For the requirements listed in exhibit 5.15, we assume that exploratory and production fields have proportionally the same areal requirements as those for oil exploration and production.

Transportation requirements for ocean transport of gas are estimated on the basis of liquefied natural gas tankers holding about 7 million ft³ of liquid and making 20 trips annually. These tankers are assumed to carry 5% of the 1 CMO/yr of gas production. In contrast to oil transportation, trucks are used to transport gas only if it has been liquefied.

---

23. Assuming that, on average, a field is depleted in 10 years, about 3,000 mi² would be added to the world's potentially contaminated area in 50 years if the production rate remained constant at 1 CMO/yr. If the demand for oil continues to grow at the current rate of 1.3%/yr, the total potentially contaminated area would expand by another 1,100 mi². Small releases of contaminants associated with transportation and refinery fires are common but ordinarily not significant.

Exhibit 5.15.

Requirements for Producing 1 CMO/yr from Gas: Selected Operations (estimates rounded)

Activity	Facility Requirements (Units)	Environmental Impacts	
		Area (mi²)	Other
Extraction			
Exploratory wells	1,700–3,500	50–100	Blowouts, flared gas, toxic drilling materials
Production wells[a]	130,000 (1,300 fields)	200	
Transportation			
LNG tankers[b]	250	20–40[c]	Leaks, fires, explosions
Pipelines	170	1,600	

[a]For 0.2 CMO/yr of nonassociated gas and 0.8 CMO from oil wells as associated gas.

[b]Carrying a total of 0.05 CMO/yr.

[c]Liquefaction and gasification facilities.

Pipelines would carry most of the gas produced. Often the gas would move by a main pipeline (typically 36-inch diameter) to a distribution point and then pass through several subsidiary pipelines before reaching homes, factories, or electric power plants. Laying down the pipelines would require the companies to acquire rights-of-way on either side of the pipeline for operational and/or safety reasons. The 1,600 mi² listed in exhibit 5.14 for pipelines accounts for this area (although we note that this land might also have limited additional uses). The gas carried by the assumed 36-inch-diameter pipelines will require a total of about 3,500 pump-compressor stations to move 1 CMO.

Requirements and Potential Impacts of Producing 1 CMO/yr from Coal

Coal can be extracted directly from outcrops, mined by stripping away an overburden (the surface material removed), removed from mine entrances, or taken from a combination of shafts and connecting passages. In the latter two cases, much additional rock is removed and generally left above ground. In surface mining, large amounts of soil may be removed to expose the coal. This soil can be stored and then replaced in a manner that restores the top (more useful) layers to productive use. Coal is also often treated with water at the mine to remove gross impurities, which are stored in waste heaps whose contents could be, but usually are not, returned to the mine.

Only a portion of the impurities associated with mined coal, including minerals and rock, are removed during washing or mechanical separation.

The organic portions of coal contain some elements (e.g., sulfur) within their molecular structures; other impurities were admixed with the organic matter when the coal was first deposited. These types of impurities cannot be removed easily by mechanical means, and they form the ash that remains after coal combustion.[24] The wash water used at the mine site is often contaminated with sulfurous acid, from the inorganic sulfides present in the coal, and/or with particulate matter, and these often flow into nearby streams. Sometimes the wash water is impounded to allow the particulate matter to settle first.

For the facility requirements shown in exhibit 5.16, coal is assumed to be produced in two types of mines:

- Western U.S. surface mines, whose coal has a heating value averaging 16 MBtu/ton and contains 1% sulfur and 12% ash
- Underground mines in the Eastern United States, whose coal has a heating value of 26 MBtu/ton, a sulfur content of 3%, and an ash content of 8%

For both types of coal, about three-fifths of what is burned is assumed to be subject to air pollution controls that require removal of sulfur dioxide by treatment with lime or limestone. The process for removing sulfur produces calcium sulfate, which is then used in gypsum boards or in making Portland cement.[25]

For a typical surface mine, the stripping ratio is often 3:1, meaning that three times as much land is removed as coal is recovered. As much as 95% of the overburden is ultimately returned to the pit created during the mining operation. (If the area surrounding the mine has little further economic value, this restoration is not likely.) The annual production from this typical mine is approximately 7.4 million tons of coal, and 1,300 mines would be required to produce 1 CMO/yr. The residues associated with that amount of coal would consist of (1) the overburden not returned to the mining pit, (2) the ash from burning the coal at power plants, and (3) the calcium sulfate product resulting from the techniques most coal-burning plants use to remove sulfur. If these residues were stacked in a 50-foot-high pile, the area covered would total between 40 and 50 mi²/CMO/yr.[26] Over 50 years, this accumulation at the current rate of coal use of approximately 0.8 CMO/yr would cover 1,600–2,000 mi² (a much larger area if the overburden is not returned).

24. Among the more hazardous compounds found in coal are, as noted, thorium and uranium, and the daughter products of their decay, which can make the stack gases of most coal-burning power plants greater sources of direct radioactive emissions than nuclear plants.

25. Although sulfur removal is growing in importance in Europe and North America, it appears to be viewed as less important elsewhere at this time.

26. For the calculations here and on the following pages, we have assumed the area is covered with a pile having an average density of 150 lb/ft³ of material.

Exhibit 5.16.

Requirements for Producing 1 CMO/yr from Coal: Selected Operations (estimates rounded)

Activity	Facility Requirements (Units)	Environmental Impacts	
		Area (mi²)	Other
Extraction			
Surface mines	1,300	34[a]	Spoil and ash run-off, sulfur and nitrogen oxides
Underground mines	2,600	31[a]	
Transportation			
Trucks (to railhead from surface mine)	300,000	?	Highway damage and diesel engine exhaust
Unit trains[b] (from underground mines to power plants)	2,600	18,000	Diesel engine exhaust

[a]The total area for the assumed mine spoil, ash disposal, and 60% of sulfur stored as calcium sulfate dihydrate. A 30-CMO coal supply (50 years at the current rate of use) would require about 1,200–1,500 mi² for the assumed mines.

[b]A train consisting of 130 cars drawn by three 3,500-horsepower locomotives on a round trip of 360 miles. One mine and unit train combination is sufficient to fuel one 900-MW_e unit at 75% capacity factor for 1 year.

A typical underground mine produces about 2.3 million ton/yr. If all of this were just coal, then 2,600 mines would be required to produce 1 CMO/yr. Each would leave about 750,000 tons of excavated materials (or the "spoil") from the mine above ground. Storage would require nearly 20 mi²/CMO of land if the spoil is piled 50 feet deep. Similarly stored ash would occupy more than 4 mi²; assuming three-fifths of the stack gases are treated, another 3 mi² would be needed for calcium sulfate. Thus, approximately 25–30 mi²/CMO/yr of surface storage would be required. At the current rate of coal use, 20–24 mi² would be covered in 1 year, and 50 years of production would require 1,000–1,200 mi² of storage space.

Review of Our Inheritance: Fossil Fuels

Our reserves and resources of conventional fossil fuels and their global distribution are summarized in exhibit 5.17. Global reserves of conventional oil

Exhibit 5.17.
Summary of Regional Conventional Fossil Resources (CMO, all data rounded)

	Reserves			Speculative (5% Chance of Recovery)		
	Petroleum	Gas	Coal	Petroleum	Gas	Coal
World Total	46	42	121	94	66	1,030
Percentage of World Total						
North America	5.0	4.4	26.7	9.8	6.9	130
Central/South merica	8.6	3.8	2.0	16	7.1	8
Europe	1.2	3.2	4.2	4.5	4.8	59
Russian Group	10.7	32.1	24.6	19	21	573
Asia/Pacific	4.9	8.2	35.3	5.1	6.5	241
Middle East/North Africa	66.3	44.8	<0.1	39	17	3
Sub-Saharan Africa	3.4	3.5	7.2	5.4	2.9	17

and natural gas are 46 and 42 CMO, respectively. These quantities will likely be exhausted in 30–60 years. If we add to these the potential quantities of oil and natural gas in our resource base, then the numbers for oil and gas rise to 140 and 108 CMO, respectively, extending the time to exhaustion by many more decades.

If we also include the unconventional resources, our total oil and gas endowments could last us a couple of centuries. In addition, our coal reserves (120 CMO) and resources (up to 1,200 CMO) could also meet our energy needs for several centuries. Although resource estimates by other analysts may differ somewhat, the main conclusions are the same. There is a large, albeit finite, supply of these fossil resources. Most of this is currently unavailable, however, which is a key point. The quantities of fossil fuels that actually become available over time will depend entirely on technical advances and our willingness to pay in monetary and environmental terms for their extraction and use. As the relatively easy-to-produce oil fields are rapidly being depleted and gas fairly rapidly depleted, we will have to rely more on oil and gas from fields that are difficult to develop.

The distribution of fossil resources around the globe is uneven. Concentrations of oil deposits in the Middle East/North Africa region, and of gas deposits there and in the Russian Group, also present a logistic challenge for the consuming nations of Europe and North America. Those latter two regions, along with major consuming nations in the Asia/Pacific region, are now the largest users of energy. Coal would seem to present fewer global energy problems, given that it is both more abundant and more widely distributed. The

Asia/Pacific region has the most coal and is where energy use is most likely to expand, whereas the Middle East/North Africa region has essentially no coal but has other resources. As discussed above, the consideration of global distribution of available resources is also countered by the knowledge that use of coal creates substantially more environmental damage than does use of oil or gas. That is to say, the existence of an available resource does not automatically mean that society considers it available for use.

Production and use of our inherited fuels require many facilities to extract the raw material, transport and refine or concentrate it to usable fuel, and then distribute the fuel to end-use facilities. Residues from the extraction of these inherited fossil resources will accumulate over time, and this impact of the use of fossil and nuclear fuels will grow. Through waste created during mining, ash residues, and solids arising from the control of its combustion products as well as production of CO_2, coal is and will continue to be the greatest environmental offender among the fossil fuels.

We are experiencing deceasing ability to access oil and natural gas and an increasing unwillingness to accept the environmental damage associated with accessing those resources. This situation indicates pursuit of a threefold strategy: (1) conserving inherited resources by changing our pattern of use; (2) increasing research and development in the geology, environmental science, and chemistry of fossil fuels to render a greater proportion of the fossil fuel endowment safely accessible; and (3) realizing that fossil resources are ultimately limited and pursuing development of our nuclear and/or income sources, such as wind and solar, to meet the increasing demand for energy.

6

Our Energy Inheritance

Nuclear Power

Of the various alternatives to fossil energy, nuclear power is the most advanced and the best positioned to become a major source of energy. It is also essentially free of CO_2 emissions, and if reducing greenhouse gas emissions is truly the highest concern, then we will have to develop nuclear power. Yet developing nuclear power would also pose challenges in terms of waste disposal, and proliferation of nuclear weapons including the risk of a terrorist organization acquiring such weapons. To some environmentalists nuclear power presents a serious, dilemma. Support or opposition to nuclear power is strongly affected by value judgments as well as lack of disseminated information on questions: What happens if there is leakage of nuclear waste someday? To what extent would people and the world be affected? Would we be trading international security for energy security—does nuclear power increase our vulnerability to terrorist attacks? The mixture of clear benefits with outstanding questions currently allows some nations to embrace nuclear power, some to accept it grudgingly, and still others choose to ignore it.

Given its availability and environmental benefits, nuclear is an option that cannot be ignored if we are to tackle the energy problem in a serious way. To assume that we can store and safeguard the waste for thousands of years may be hubris, but we come out in favor of developing nuclear technology in part because we already have to store the legacy nuclear waste that has been generated over the last 50 years. Another 60 or so years of waste will represent a marginal addition to that enormous task, but it would buy us badly needed time to carefully develop other energy sources that do not entail net greenhouse gas emissions. Also, we find that many of the concerns raised against the development of nuclear power are vastly exaggerated. For example, as we describe in this chapter, safe storage of the waste does not require 10,000 years: if we use reprocessing technologies, the remaining waste could be rendered benign in a couple of centuries. We refer readers seeking to place the risks of nuclear power into perspective to a highly readable and informative book *Physics for Future Presidents,* by Richard Muller, a professor at the University of California–Berkeley, with first-hand experience in national security measures.[1]

Overview

Nuclear energy has a relatively short history. It was not until the late 1930s that scientists realized the vast amount of energy locked in the nucleus of the atom. Much of the initial effort was focused on producing a nuclear weapon because of the overwhelming military advantage that it could provide. It was only after the Second World War that nuclear energy was developed for the purpose of producing electrical power, and the first commercial nuclear plant was established only in 1960.

Between 1960 and 1987, the global installed capacity for nuclear power grew from less than 1 GW to more than 350 GW, reflecting an average growth rate of more than 25% per year. In 2006, about one-sixth of the world's electricity was produced by nuclear reactors. This rapid rise in nuclear power demonstrates that under certain circumstances a new source of energy can be developed very quickly, and it behooves us to understand those circumstances as we seek alternate sources of energy to match the rate of increase in worldwide demand for power. In this history, extraordinary support by the military of many nations was a big factor in development of nuclear power. Nuclear power was further enabled by the fact that the infrastructure for the transmission and distribution of its product, electricity, was already in place. When demand for electricity increased in the decades following the Second World War, nuclear energy was readily added to the mix of power generation systems. While these exact circumstances are not likely to be repeated as we attempt to promote new energy technologies, they suggest that appropriate actions by the government could enhance chances of successfully deploying these technologies.

In this chapter we begin with a review of nuclear power and some of the underlying physics. We then examine how power is generated in a nuclear plant by following what happens to the fuel from mining to final disposal. We discuss nuclear power generation in some detail, because nuclear power technology is both desirably scalable and undesirably misunderstood. To facilitate this discussion, we use the example of a common reactor technology, the light

1. Richard Muller. *Physics for Future Presidents,* Norton, New York, 2008.

water reactor (LWR). This discussion also brings to light some of the issues and concerns, and subsequently we examine how alternate designs can deal with them and what trade-offs would be involved.

As for all energy sources in this book, we then discuss our reserves and resources. It becomes evident that the availability of nuclear reserves and resources will be strongly influenced by the choice of reactor technology and fuel processing methods. These choices also affect the greenhouse gas emissions associated with preparation of the fuel and construction and decommissioning of nuclear power plants. Finally, we address the debated drawbacks of nuclear technology, which concern the health and safety of the populations around the nuclear facility, the proliferation of nuclear weapons, and the long-term storage of the radioactive spent fuel. Some of these drawbacks are simply fear spawned from misinformation, some require political solutions, and others require technological solutions.

Basic Physics of Nuclear Processes

The process of producing electricity from nuclear energy is in some ways much like that involved in the fossil-fuel plants discussed in chapter 3. Nuclear fuel is consumed to produce heat. The process is often referred to as the fuel being "burned."[2] This heat is used to produce a heated fluid, typically steam, which is used to turn electric turbines that generate electricity. Where nuclear power plants differ markedly in practices from fossil fuel plants is in the standards and safeguards for plant construction and in the disposition of the fuel once it is exhausted (referred to as "spent").

To appreciate the similarities and dissimilarities between chemical and nuclear processes, it is appropriate to consider the structure of an atom. An atom consists of a tiny positively charged nucleus surrounded by a cloud of negatively charged electrons. Most of the mass of the atom is in the nucleus, which itself consists of positively charged protons (p) and electrically neutral particles, the neutrons (n). The number of electrons surrounding the atom equals the number of protons, and these particles give rise to the chemical identity and properties of the atom. The number of neutrons can vary, giving rise to different isotopes of the element. Even though the chemical properties of different isotopes of the same element are essentially identical, their nuclear stability and interaction with other neutrons can be vastly different. That is why, in the context of nuclear energy, it is important to specify the particular isotopes of various elements.

---

2. We normally think of burning as the process in which materials like wood or coal combine with oxygen to produce heat (i.e., combustion), but "burning" is also used to describe the process of consuming nuclear fuels.

Hydrogen, symbolized as ^1H, is the simplest atom and has one proton and one electron. Deuterium is an isotope of hydrogen, having a neutron in addition to the proton and the electron. Deuterium weighs twice as much as hydrogen, and its symbol is D or ^2H.[3] Tritium (T or ^3H) is another isotope of hydrogen and contains two neutrons, plus the proton and the electron. After hydrogen, the next element, helium, has two protons and two electrons. Most commonly its nucleus also has two neutrons, and that gives rise to the ^4He isotope. The nucleus of ^4He is also referred to as an alpha particle and is emitted during the radioactive decay of many heavy atoms. An isotope of helium with only one neutron, ^3He, is also important in many nuclear reactions.

At the other end of the periodic table, we have the heavy atoms. Uranium is a fairly common heavy atom and also the one that we are most concerned with in this chapter due to its use in nuclear power production. The three naturally occurring isotopes of uranium are ^{233}U, ^{235}U, and ^{238}U. Most nuclear plants use ^{235}U as fuel. Thorium, another heavy element, has a single stable isotope, ^{228}Th, and could someday play an important role in nuclear power production.

The interactions between protons and neutrons generate attractive and repulsive forces, and stable isotopes have nuclei in which these forces are balanced. In a few isotopes of naturally occurring elements this balance can become perturbed, often spontaneously by chance, and the atomic nucleus expels a small particle or high-energy radiation—a gamma ray. For example, potassium is a ubiquitous element naturally present in seawater and most minerals. It is also vital for proper functioning of the cells in our bodies. Potassium has an isotope with a mass of 40 units (^{40}K), occurring in nature with an abundance of 0.01%. ^{40}K occasionally emits an electron (also called a beta particle) from its nucleus and transforms into an isotope of calcium, ^{40}Ca. This happens only infrequently, and it would take about 1.3 billion years (the half-life of this process) to convert half the atoms in a sample of ^{40}K into ^{40}Ca. Naturally occurring uranium isotopes, ^{235}U and ^{238}U, have half-lives of 0.7 and 4.5 billion years, respectively. However, if bombarded by neutrons, a heavy element with an atomic number greater than 90, such as uranium, can capture neutrons to create a very unstable nucleus that then decays by breaking into smaller nuclei in a fraction of second. This process of breaking apart into smaller nuclei is referred to as nuclear *fission*. In contrast, the process of melding smaller nuclei into a larger nucleus is referred to as *fusion*.

In both fossil fuel and nuclear fuel consumption, regrouping reactions result in formation of products that are more stable and contain less energy than the starting materials, and this difference in energy between starting materials and products is released as heat. The energy released during combustion of fossil fuels comes from the regrouping of the atoms of fuel

---

3. The superscript preceding the elemental symbol refers to its atomic mass. Thus, deuterium has an atomic mass of 2.

molecules and oxygen. The numbers of carbon, hydrogen, oxygen, and other atoms remain the same during the process, but how they are combined with one another changes. In a nuclear reaction, the atoms themselves change, but the sum total of protons and neutrons that make up the nuclei of atoms remains the same.

Splitting, or fission, of atoms heavier than iron into smaller atoms produces a more stable configuration, while fusion of lighter atoms generally leads to more stable configurations. Formation of more stable products manifests itself in a change in mass, and this is a key point. Taken together the product particles are slightly, but significantly, of lower mass than the reactants. The energy (E) released in the process is related to this difference in mass (m), in accordance with Einstein's celebrated equation, $E = mc^2$, where c is the speed of light. This energy is what keeps the positively charged protons bound together in such close proximity in the nucleus of the atom, instead of repelling each other and flying apart.

A marked difference between chemical and nuclear reactions is that nuclear reactions are associated with energy releases that are about *ten million times larger* than for chemical reactions. Thus, about 2,000 tons of ^{235}U can release as much energy as burning 4.2 billion tons of oil (which is 1 CMO).

By virtue of this large energy release, nuclear reactors require much smaller amounts of material for fuel and tend to have substantially smaller footprint than do other power sources. A 2.0 GW reactor can be fenced within a 60-acre plot, about one-tenth of a square mile. Compare that with as much as a square mile needed for a comparable coal plant with its stores of coal and ash, and about 6 mi^2 for a solar thermal facility. For a given power capacity, the quantity of materials required for constructing a nuclear plant is also less than required for a coal or wind power facility.

We now turn to the specific nuclear reactions that are relevant to nuclear power generation.

Fission

Currently all nuclear power plants use fission of uranium as the primary source of energy. When a neutron adds to a ^{235}U atom, it results in ^{236}U, which is extremely unstable and immediately breaks apart into atoms of lighter elements. The most common of these fission reactions produces an atom each of barium and krypton along with three neutrons:

$$^{235}U + n \rightarrow [^{236}U] \rightarrow {}^{144}Ba + {}^{89}Kr + 3n$$

The nuclei of barium and krypton contain 56 and 36 protons, respectively, and together these equal the number of protons in the uranium nucleus, 92. The number of neutrons is also preserved in the process, but as mentioned above, the combined mass of all particles in the right-hand side of the equation is slightly less than the combined mass of the starting ^{235}U and neutron.

As per Einstein's equation, some of the mass of the starting materials has been converted into energy. This mass difference manifests first as kinetic energy in the fast moving fragments, and later as heat as these fragments collide with other atoms and slow down.[4]

The fission of ^{235}U by bombardment with a neutron produces more neutrons (called daughter neutrons), which in turn could induce fission of additional atoms. The key for sustained release of energy is that ^{235}U, or some other atom that can undergo fission (i.e., a fissile atom), captures at least one of these daughter neutrons and undergoes fission, producing more daughter neutrons and continuing the process in what is called a chain reaction. The neutrons produced during the fission are very fast, moving at speeds of 11 million miles per hour, and the probability of their reaction—or cross section, as it is referred to—with the relatively rare ^{235}U is very small. There is a more abundant isotope, ^{238}U, which is more likely to capture the fast neutrons. ^{238}U is not fissile; it does not break apart into smaller nucleic releasing energy and producing daughter neutrons. This capture of neutrons by ^{238}U is undesirable on several counts. First, it interferes with the chain reaction. Second, and perhaps more important, the capture produces ^{239}U, which through a succession of radioactive processes transforms in a matter of minutes to an isotope of plutonium, ^{239}Pu. ^{239}Pu is radioactive (emits alpha particles), but it decays slowly, with a half-life of more than 24,000 years. It is also toxic, because it is preferentially absorbed by the bone, and although its alpha particle radiation does not penetrate very far, it can still affect the marrow and cause leukemia. The radioactivity and very long life of some of its decay products mean that ^{239}Pu and other reaction products must be handled with special precautions. Thus, untreated spent fuel has to be safeguarded for a very long time—longer than the history of human civilization! Finally, ^{239}Pu is also extremely fissile and has been used to make nuclear weapons, which creates most of the issues surrounding nuclear proliferation. We discuss the issues of radiotoxicity of spent fuel[5] and nuclear proliferation later in this chapter. In the present discussion, it is notable that a significant contributor to the storage issues currently associated with nuclear power arises from a side reaction that does not particularly support the power-generating process.

To minimize the formation of ^{239}Pu, the initially formed daughter neutrons must be slowed down ("moderated" and "thermalized" are two other terms used to describe this slowing down) from speeds of about 1,000,000 mph to 2,200 mph. Slowed down, the probability of their reaction with ^{235}U is sufficiently high to produce another fission and thus continue

---

4. In this example, both the number of protons and the number of neutrons are conserved. There are also nuclear reactions in which neutrons are transformed into protons and vice versa. Capture of a neutron by ^{238}U to produce ^{239}Pu is one such example.

5. As far as radiotoxicity of spent fuel is concerned, the danger from some of the fission products, such as strontium-90 (^{90}Sr), is even greater than that of ^{239}Pu.

the chain reaction. Favoring the fission of ^{235}U over capture by ^{238}U is another reason why most nuclear reactors use uranium that has been enriched in ^{235}U, from the 0.7% natural abundance to between 3% and 6%. Most nuclear power reactors also employ moderators such as water and graphite to slow down the neutrons. The situation where daughter neutrons from each fission event lead to one additional fission event is also called the critical condition, and maintaining this condition depends on many factors, such as the level of ^{235}U enrichment, the amount of fissile material, and the presence of moderators. Operators of nuclear power plants constantly adjust the flux of neutrons within the reactor to ensure that the criticality is maintained.

Once the process of fission had been discovered, scientists immediately realized that if each fission event leads instead to more than one fission event (i.e., a supercritical condition), the situation could quickly escalate and result in a massive explosion. Having too high a concentration of fissile atoms can cause supercriticality, which is why weapons-grade uranium normally has ^{235}U enriched to around 80% or more, compared to the 3–6% level of enrichment used for power production.

Fusion

In the case of light elements like hydrogen and helium, the process associated with the release of energy is fusion, which is the opposite of fission. In fusion, the heavier atom product is in a lower energy state compared to the lighter starting materials. An example of this process is the fusion of two nuclei of deuterium (^{2}H), also known as deuterons, to give either ^{3}He and a neutron or ^{3}H (tritium) and a proton. Nuclear fusion is what powers the sun. Nuclear fusion also powers the H-bomb, which takes an explosion from a plutonium bomb to trigger the fusion of hydrogen. Although we can release energy by fusion in the H-bomb, releasing fusion energy in a slow and sustained manner for power production has thus far eluded us.

In order to fuse the atoms, we have to overcome strong repulsive forces, first of the electron clouds around the nuclei and then between the positively charged nuclei. Having only one proton in their nuclei, hydrogen atoms offer the least repulsive force to overcome. Hence, most attempts at producing fusion energy have focused on the isotopes of hydrogen. Initiating fusion reactions generally requires temperatures in excess of a million degrees. At these high temperatures, matter exists in what is known as a plasma state, where the electrons are stripped off the atoms, and bare nuclei are made to collide with each other at high speeds. Of course, if fusion occurs, the vast amounts of energy released could be used to promote fusion of other nuclei as long as they are confined within the hot zone. Confining these hot nuclei in a small volume generally requires a high vacuum system, but fast neutrons produced in the fusion process pierce holes in the vessel holding the vacuum. For these and many other reasons, it has not been possible so far to sustain a fusion process by this approach.

Scientists have also attempted to effect controlled release of energy from fusion by focusing lasers that heat tiny volumes of hydrogen and tritium to the required high temperatures. However, all attempts to sustain a fusion process have so far required more energy in the form of lasers or particle accelerators to start the process than was released by the achieved fusion.

In a complete departure from the traditional approaches to effect fusion, in 1989 Martin Fleischmann and Stanley Pons[6] reported observing sudden large energy releases during prolonged electrolysis of heavy water with palladium electrodes. The initial explanation was that once the concentration of the deuterium atoms in the palladium built up to very high levels during electrolysis, the lattice forces squeezed them to give rise to the observed "cold fusion," in contrast to the "hot fusion" that scientists have been experimenting with using plasmas and lasers. The report by Fleischmann and Pons engendered much interest and controversy, and many other researchers attempted to reproduce the results. Most were unsuccessful, as were the attempts to monitor the formation of expected products such as ^{3}He. However, a few groups confirmed the observation of sudden releases of very large amounts of energy that was outside the range for purely chemical reactions. Even today, it is not clear what causes the occasional bursts of energy during this electrolysis.

Because many attempts at reproducing the results failed, and because no satisfactory theory that could explain those observations was found, research on cold fusion became stigmatized. Researchers investigating it found themselves ostracized, and only a few have continued. Nonetheless, the promise that this approach holds is so vast that it cannot be ignored, and research in this area should be supported. Indeed, some work on cold fusion is continuing, now under a more accurate name of solid-state nuclear phenomena. If controlled fusion were to be realized, it would certainly solve our energy problem. However, even after it has been discerned and demonstrated, it would take several decades for the technology to be adopted widely and have an impact on our energy consumption. For now, we are limited to what is available to us.

The Technology of Nuclear Power Generation

The foregoing discussion described how nuclear processes can produce large amounts of energy in the form of heat. There are many different ways to

---

6. M. Fleischmann and S. Pons, "Preliminary note: Electrochemically induced nuclear fusion of deuterium," *Journal of Electroanalytical Chemistry,* vol. 261, issue 2 (part 1): 301–308, 1989.

conduct these processes and many applications to harness the energy released in the reactor. Most reactors convert the heat to produce electric power. One other application that could use large quantities of energy generated in nuclear reactors is desalination of seawater. This application could become important as we face shortages of potable fresh water, the production of which often requires expenditures of large amounts of energy. There are also small nuclear reactors used to prepare isotopes for medical and diagnostic applications.

In this chapter we focus on nuclear power production using the light water reactor (LWR) to step through the process from mining the ore to final disposition of the spent fuel. In the current practice, which is also referred to as the once-through cycle, uranium goes successively through stages of mining, milling, enriching, power production, interim storage at the plant, and finally to the geologic repository. Following discussion of that process, we look at alternate designs that are in use or have been proposed to address some of the specific issues with LWRs.

The once-through fuel cycle is relatively simple to execute but is wasteful. It not only produces a large quantity of nuclear waste but also converts less than 10% of the energy contained in the uranium to power, the rest being rejected in the fissile products waste stream. There are alternate schemes in which the fuel from interim storage is reprocessed to varying degrees. If adopted, these schemes would extend the life of uranium reserves, reduce the volume of the waste, and also shorten the time period for the radiotoxicity of the waste to abate to the level of natural uranium (mineral ore).

Uranium Mining and Milling

Uranium is a fairly common element in Earth's crust, slightly more abundant than silver. Pitchblende (or uraninite) is the important oxide mineral of uranium and consists mostly of uranium oxide (UO_2). Monazite sands, which are phosphate minerals rich in thorium, also contain uranium. Because uranium oxide is easily dissolved by groundwater and also readily precipitated by slight changes in acidity, it migrates through aquifers and is found in small quantities in association with many other rocks such as granite and coal deposits. There are large deposits of uranium in North America, sub-Saharan Africa, and Asia/Pacific regions (especially in Australia) and in the Russian group.

Uranium is usually mined in locations where its concentration ranges from 0.05% to 5.0%. Land-based deposits with uranium content as low as 0.01% have also been exploited. As the concentration of uranium in deposits decreases, it gets progressively more expensive to produce fuel. Uranium is often associated with phosphate and gold deposits and can be obtained from processing their mining residues. Seawater and ocean sediments also contain low concentrations of uranium that could be extracted and used, although currently those resources are not economically competitive with other sources.

Uranium ores occur in underground formations as well as near the surface, and techniques used to mine coal are also applicable here. About 40% of the uranium is extracted by underground mining, and about 30% by open pit mining. Currently, for equal amounts of energy that is produced from the fuel, the quantity of earth moved in surface mining of uranium is roughly the same as that for coal—about 3 tons of earth for every ton of ore. However, unlike coal, uranium can also be extracted by "solution mining," in which chemical reagents are piped into the deposit to dissolve the uranium. The resulting solution then flows through pipes from the underground deposit to the surface. Recovery in this way is not nearly as disruptive of the environment as is conventional mining, and it requires less land. Around 20% of uranium is mined by this solution method. The remaining 10% of uranium is obtained as a by-product from gold and phosphate mining operations.

Mining uranium ores brings the uranium and its associated radioactive daughter products to the surface. The more valuable portions of the mine products are shipped to mills, where relatively pure uranium compounds are produced. Materials with lower concentrations of uranium and decay products are left on the surface at the mine site. If material left on the surface is exposed to erosion by water, the daughter products can be distributed to streams and thus contaminate a considerably larger area.

Raw uranium ore is usually transported by truck to mills that concentrate the uranium. The milling process essentially consists of crushing the ore and treating it with acid to dissolve the uranium and separate it from dirt and rocks, followed by neutralizing the acid to precipitate the uranium as a solid oxide, U_3O_8. This precipitate is also referred to as the yellow cake, which gets shipped to facilities that further process it to make fuel for nuclear reactors.

The wastes from the concentration step include trace amounts of uranium and almost all of the uranium daughter products. These are carried to waste "heaps" where, in U.S. practice, they are generally placed atop water-impenetrable barriers and covered with similar barriers and then soil. The rainwater runoff from the discarded materials is monitored, and any uranium and its daughter products dissolved and carried from the heap by the rainwater are recovered. During these operations, miners and mill operators are exposed to radioactive materials, particularly radon. (The same is true for many other miners and for coal and ore handlers, but the radon concentrations they encounter are lower.)

Early uranium mining and milling operations were carried out without much regard to potential health hazards. Estimates of annual fatalities and serious injuries for workers involved in underground and open pit mining were 0.1% and 0.06%, respectively, and were 0.013% for milling operators. In fact, the known mortality rate—mainly from lung cancer but also from other malignant growths, including those of the liver—for uranium miners working in East Germany in the late 1940s has been more than 50%. Unlike the current situation, these underground miners operated in poorly ventilated mines, did not use masks, and were probably smokers—all factors that increase the probability of cancer. Similar but less severe effects were noted among uranium miners working in the western United States during the same period. In recent years more efforts have been made to protect uranium miners working in both underground and surface mines, which account for more than 70% of the uranium mined. As noted above, the remaining uranium is obtained from solution mining, from leaching waste heaps at gold mines, and from phosphate concentrating operations, where the exposure hazards are considerably lower.

Now that radiation hazards are recognized and steps have been undertaken to protect the workers, the injury and fatality rates have fallen substantially. According to the World Information Service on Energy (WISE) Uranium Project that collected health data from the surface and underground operations at the Jabiluka mining operations in Australia,[7] the main source of radiation exposure is the radon gas that workers inhale. Radioactive decay of radon leaves solid material deposits that are also radioactive in the lungs. This process accounts for 69% of the total radiation exposure for underground and 30% for aboveground miners. The current mines are well ventilated so that radon does not accumulate in the working areas, and the workers wear masks to minimize exposure to particulate matter. The radiation emitted by materials in the ore body (the uranium and all of its daughter products) account for 28% and 60%, respectively, of exposure for the underground and surface workers.

The WISE analysts extended the data to the approximately 260,000 underground and 2,500 aboveground (open pit) miners now working worldwide, and estimated 44 additional cancer deaths per year, or about 0.73 deaths per year per 1,000 tons of uranium mined in current operations. To put the level of exposure in perspective, the U.N. Scientific Committee has placed the annual effective radiation exposure from worldwide nuclear operations at about one-half that from all types of mining worldwide (discussed below under Concerns About Nuclear Power-), and about twice that encountered by air crews—who spend much of their time flying at a high altitude of 30,000 feet where they are less shielded from cosmic radiation than the general population.[8] This comparison would suggest a similar casualty rate from radiation for workers mining uranium as for people mining other materials or working in airplanes.

Uranium Enrichment

The yellow cake produced at the end of the milling process contains the three isotopes of uranium, (^{233}U, ^{235}U, and ^{238}U) in their natural abundances of 0.004%, 0.7%, and 99%, respectively. LWRs require the concentration of

7. World Information Service on Energy (WISE) Uranium Project. *Radiation Exposure for Uranium Industry Workers,* October 2006. http://www.wise-uranium.org/ruxfw.html. Accessed August 2007

8. United Nations Scientific Committee. *Sources and Effects of Ionizing Radiation,* 2000.

^{235}U to be increased to about the 4–5% level. This enrichment is commonly achieved by gas diffusion, for which the solid oxide of uranium has to be converted into a gas. The gas commonly employed for this purpose is UF_6, which is extremely corrosive and would attack most materials. The invention of Teflon in 1938 happened just in time, and it was used to coat the pipes and vessels needed for enriching uranium for the Manhattan project. Enrichment takes place because the slightly lighter $^{235}UF_6$ diffuses out of tiny pores slightly faster than does $^{238}UF_6$. By taking the enriched stream from one diffuser into a second diffuser, and repeating the process many times, the abundance of ^{235}U is increased to desired levels for nuclear reactors, or even to the much higher levels required for making weapons. The enriched gas is then converted back into uranium oxide for packing into rods that will be used in the nuclear reactor.

As can be gathered from this discussion, enrichment is a tedious and energy-intensive process. A significant improvement in the process involved the introduction of the ultracentrifuge, whose use requires less energy. The yellow cake material is first converted to gaseous UF_6, which is then spun at very high speeds in the ultracentrifuge. Centrifugal force preferentially pushes the heavier isotope away from the center of the spinning axis. As in the case of separation by diffusion, achieving the final enrichment requires repeatedly performing the ultracentrifugation. The centrifuge separation is not simple. Any material failure or shrapnel from breakage could lead to a cascade of failures. Therefore, the centrifuges are constructed from special materials made to handle the stress from spinning at very high speeds. The alignments of each of about a thousand centrifuge tubes—and the rate of flow between them—are precisely controlled. Even with this high level of effort, the ultracentrifugation process is the preferred method of enrichment because it is faster and also requires about 50-fold less energy than does the diffusion process. To produce the 4 to 5% or so of enriched uranium needed for power plant reactors requires about half the energy needed to produce the highly enriched uranium that is suitable for nuclear weapons—thus the governmental and public concern when nations in politically unstable regions enter into centrifuge-based isotope separation, even though it is presumably intended for power production.

Fuel Rod Assembly

Following enrichment, the uranium hexafluoride is converted to the oxide. The enriched UO_2 is formed into pellets that are typically 0.3 inch in diameter and 0.5 inch long. The pellets are placed in tubes that are about 12 feet long and made of an alloy of zirconium and tin (zircaloy). This alloy is chosen because its constituent metals do not absorb neutrons very efficiently. It is operationally advantageous to ensure that heat generation is uniform across the length and width of the reactor core, and sometimes rods with pellets containing neutron absorbers, or "poisons," are used to manage the neutron flux across the reactor to obtain optimum power and uranium utilization.

At the start of reactor operations, as many as 220–270 fuel-bearing rods are placed into assemblies that may also contain monitoring instruments and neutron-absorbing control rods. With the control rods in position, the system is subcritical and a nuclear chain reaction cannot take place. About 100–200 such assemblies are lowered into the water tank of a typical LWR. Once the tubes are in place, the control rods are gradually withdrawn to increase the neutron flux and achieve criticality. In some instances, the fuel is "ignited" by introducing a small quantity of highly radioactive material that provides a source of neutrons. A mixture of plutonium and beryllium is a commonly used igniter for inducing fission in some of the fuel pellets to start the whole process.

As the nuclear reaction proceeds, the water in the tank gets heated. This water is the "coolant," the medium that transfers heat from the nuclear reactor to the power generator. What happens next depends on the kind of LWR. In boiling water reactors (BWR), the water is allowed to boil, and the resultant steam is used to drive a generator. Since this water is in direct contact with the reactor core, any leak in a BWR reactor could result in release of radioactivity through the steam cycle, although thus far such has not been a significant problem with BWR operations. In pressurized water reactors (PWR), the water in contact with the nuclear fuel is contained in a pressure-tight vessel, and the fission heat is extracted by circulating a separate loop of water that never comes in direct contact with the nuclear fuel. Because of the high pressures, these reactors are built with an extra pressure-containing structure, in addition to the concrete containment that is required for both PWR and BWR reactors.

As noted, the fuel rods also contain "burnable" poisons; these are elements that capture some of the neutrons produced during fission, thus reducing the initial power generated both within the individual pellets and within the reactor. This capture is necessary to avoid overheating in certain regions of reactor, while at the same time allowing maximum heat output over time from the reactor as a whole. Without these poisons and those contained in the reactor's "control rods," power production would be highest at the center of the cylindrically shaped reactor core and lowest at its outside, top, and bottom. As a result, the fuel in the center region would soon be "spent" or at least relatively unproductive compared with the fuel in other regions.

Besides functioning as the coolant, water also acts as a moderator of the fast neutrons, and it performs the role of regulating the reaction during operation. Its effectiveness as a moderator depends on its density. Should reactor temperature start to rise from local heating, the density of water decreases, and fewer fast neutrons are thermalized. As a result, the fission in the fuel slows down. This negative feedback helps maintain stability during operations.

In the early days of reactor operations, the zircaloy covering tended to corrode and fail, resulting in frequent reactor shutdowns to replace fuel. Subsequent improvements in the properties and corrosion resistance of the zircaloy covering and in the production of more uniform UO_2 pellets increased the reliability of individual rods and enabled better power regulation. These improvements in fuel longevity have increased the time between reactor shutdowns, and with much better management of "outage" activities, U.S. reactors now produce at their full power output rating about 90% of the time.

With the buildup of fission products, the chain reaction becomes more difficult to sustain, and the fuels rods have to be replaced. The spent rods still contain a large amount of the enriched ^{235}U originally in them. They also contain many of the actinide elements that were formed by the capture processes as well as the highly radioactive fission products, and they are also literally hot. At the termination of power production in a nuclear reactor, about 12% of the heat being generated is coming from the decay of these fission products. The spent fuel rods are withdrawn from the reactor and placed in pools of water to let the highly radioactive components decay down. After several years, the rods are "cool" enough that they may be retrieved from the pool for reprocessing, as is done in many countries such as France, the United Kingdom, and Japan. In the United States, where spent fuel reprocessing was stopped in 1974, the rods are either left under water or stored in dry, air-cooled concrete structures awaiting their final disposal underground in yet-to-be-determined deep geologic sites. In countries that practice reprocessing, the uranium and plutonium is chemically extracted from the spent rod for use as fuel, and the remaining materials are converted into a glass—vitrified—and encased in steel containers.

Although the rate of heat release from spent fuel declines rapidly, it also necessitates the availability of a cooling source at any reactor shutdown. The requirement that cooling water be readily available is particularly severe in the case of an accident. This requirement and its ramifications are discussed below in the context of the Chernobyl accident.

Other Reactor Designs

Only a few reactor concepts have been commercialized since the mid-1940s, although many more were, and are still being, suggested. The variations in designs stem from the use of different moderators and coolants. As described above, most of the reactors use ^{235}U as the fuel. Since it undergoes fission with thermalized neutrons, these reactors use moderators to slow down the initially produced fast neutrons. Light (normal) water is a reasonably good moderator, but its effectiveness is limited by the fact that it can also capture the thermalized neutrons. Thus, LWRs require enrichment of ^{235}U to compensate for the decreased availability of slow neutrons. Heavy water is a much more effective moderator, and it does not capture the thermalized neutrons. Thus, it can sustain a chain reaction without enrichment of the uranium fuel. Of course, it does require enriching light water, H_2O, to pure D_2O. Because D_2O water is 11% heavier than light water, this enrichment requires much less effort than enrichment of $^{235}UF_6$ ($^{238}UF_6$ is only 0.3% heavier than $^{235}UF_6$). Reactors using heavy water as the coolant and moderator are also referred to as

CANDU reactors. The name reflects their Canadian design (CAN) and the use of deuterium and uranium as the chief elements (DU). It also reflects the can-do spirit of the team that invented it.

Other design variations include the use of graphite as the moderator. LWRs use graphite rods to control the reactor by adjusting the neutron flux, but water is the main moderator in those reactions. Reaction of superheated water with metals can be hazardous under some circumstances. Use of graphite as the moderator allows the use of fluids other than water as the coolant. Thus, there are reactors with graphite moderators and either water or gases such as CO_2 and helium as coolants.

In the reactors that use thermalized neutrons for engendering fission, the buildup of fission products in the nuclear fuel adversely affects the overall efficiency because many of them have a high capture efficiency for neutrons. The fast breeder reactors (FBRs) ameliorate this problem by starting with a higher level of enrichment of ^{235}U, around 20% instead of 4%, or with reactor-produced plutonium, and surrounding the active core of the reactor with ordinary uranium, depleted uranium, or thorium. These elements capture the fast neutrons and produce more fissile fuel that can be later extracted from the spent fuel. By starting with a higher degree of enrichment, the breeder reactors use the ^{235}U fuel much more efficiently and to a higher degree. Because of the low rejection of ^{235}U in the spent fuel, breeder reactors consume only about one-tenth of the uranium that LWRs use for comparable power output and also produce about one-tenth of the waste. The core of the FBRs also operates at higher temperatures and therefore requires more efficient coolants such as liquid sodium or highly refractory organic liquids.

The preferred coolant for FBRs is sodium because, unlike water, it does not markedly affect or moderate the neutron spectrum. Only two reactors of this type are in commercial operation, in Russia and in France. The French reactor has encountered difficulties, primarily in the coolant system, that have led to delays in its development program. We do not know about the experience at the Russian breeder. The United States has a long history of experimentation with sodium-cooled breeder reactors. In fact, the first nuclear power generation in the United States was achieved with an experimental breeder in 1951, the EBR-I, which was followed by EBR-II that was coupled to an experimental fuel-recycling program. Difficulties with this program and the general decision by the U.S. government to concentrate on conventional reactor programs led to canceling of breeder reactor development activities in the United States. A "full-scale" breeder reactor (only 61 MW) was operated very briefly by the Detroit Edison Company in 1966. This reactor was shut down shortly after full power operation began because a part of a flow baffle broke off and clogged a cooling channel and resulted in fuel melting. This accident, added to the previous design and fuel manufacturing problems that reduced the potential power output, led to a decision to permanently close the reactor. It was subsequently dismantled, and the components were safely stored. No significant effects were noted off-site because of this accident.

The Next Generation of Nuclear Reactors

The new reactors that will be built for utility service over the next decade and beyond are expected to be improved versions of the boiling and pressurized types in use now and will also entail some new concepts. Often referred to as generation IV (or Gen IV) reactors, they will range in size from about 600 to 1,500 MW$_e$ and will be designed and built with improved safety features that provide greater assurance of effective heat removal in the event of a loss-of-coolant accident. The larger size of some of these plants makes the expenditures for installing extra safety features such as dual 50-inch-thick concrete walls and passive systems to turn off the reactor more affordable as a result of the larger revenues. In previous designs, emergency shutdown of the system required very large diesel pumps and other electronically actuated systems that occasionally failed during testing; Gen IV reactors will rely on gravity flow of stored water or expansion of compressed air for these functions and will require constant human intervention to keep the reactor operating. In the absence of commands from an operator, the system will automatically and safely shut itself down.

The new designs have also benefited from analysis of the behavior of radioisotopes following the Three Mile Island accident. This analysis demonstrated that, contrary to earlier speculation, iodine and other semivolatile materials did not escape the containment even though there was extensive melting of the fuel elements and large quantities of the more volatile materials in the fuel were released. With passive systems for cooling, reactor designers are now confident that large amounts of radioactive material will not be released if the containment is at least as strong as the one used at the Three Mile Island reactor. Some designs provide that in the event of an accident, all that the workers have to do is leave and the reactor will safely shut itself down. In addition, their simplified designs should reduce field construction times, as will the greater use of factory-built modules that can be transported preassembled to a site instead of having to be assembled at the site (which is a more time-consuming activity). Some reduction in capital cost per MW will also be achieved through the use of larger units.

Graphite-Moderated and Helium-Cooled Reactors

South Africa and China are developing graphite-moderated and helium-cooled reactors, and a similar development program has been proposed in the United States. Such a reactor can operate at higher temperatures and can be used with a containment structure that is less massive—and thus less costly—than those currently in operation. The high temperature of the exit gas, as much as 1,650°F (900°C), makes the use of gas turbines with thermal efficiency as high as 50% feasible. The heated helium can be used to convert water to steam and thus to generate electricity in a turbine, or the helium itself—which is at a higher temperature—can be used to drive a gas turbine, resulting in higher conversion efficiency. The higher temperatures can also be used for more efficient production of synthesis gas from coal, with a savings of about one-third of the energy (and thus coal consumption) normally required. The higher temperatures also make the production of hydrogen gas from water by thermal means technically feasible.[9]

The fuel in the new reactor designs consists of spherical "pebbles" about 2 inches in diameter. Each pebble contains about 1,500 microspheres, each of which has a slightly enriched UO_2 core. The core is surrounded by successive layers of graphite and other carbonaceous materials for moderating the neutrons. The microspheres are compressed with additional graphite to form the pebble. Having graphite moderator intimately connected to the reactive uranium precludes meltdown of the core that could happen in case of accidental loss of water in an LWR. The reactor admits the pebbles into the top of a vertical cylinder and lets them pass, as a bed, downward against an up-flowing helium stream. Control rods in the cylinder moderate the pebbles' fission heat release. Once the pebbles reach the bottom of the vertical shaft, they are examined to determine their degree of enrichment and then either recycled or discarded.

Although this reactor avoids the hazards of nuclear meltdown implicit in water cooling, its safety has been questioned because of the potential for air-graphite burning that could occur if a loss of the helium coolant occurred. However, it is unlikely that sufficient oxygen can enter the high-temperature core (>1,000°C) against the helium outflow to initiate burning. Safe use would be achieved primarily by conservative design of the reactor and the containment building.[10]

Other high-temperature graphite moderators that have been proposed would use solid blocks of graphite as moderators surrounding fuel elements; those elements would be the same as the pebble-bed type but designed to operate longer at even higher temperatures.

Alternative Water-Cooled Reactors
Several reactors under development are improved versions (i.e., less prone to be destroyed in case of an accident) of existing BWR and PWR designs. The first of these, which went into service in 2007, was the 1,350-MW advanced BWR (ABWR) developed jointly by GE and Japanese-based organizations, and commercialized by GE Hitachi Nuclear Energy. Because the advances in this ABWR are incremental in nature and relate mainly to preventing loss of cooling water, it is also characterized as a third generation (Gen-III) reactor. The

---

9. The effects of high temperatures and the often-corrosive environment of these processes make many of the projected achievements problematic.

10. Near Term Deployment Group, U.S. Department of Energy. *A Roadmap to Deploy New Nuclear Power Plants in the United States by 2010,* vol. 1, Summary Report, U.S. Department of Energy, Washington, D.C., 2001, as quoted in D. Bodansky, *Nuclear Energy, Principles, Practices and Prospects,* 2nd ed., Springer, New York, 2004.

ABWR's simplified design incorporated features that reduced construction time in Japan from the usual 4 years to 3 years for two reactors at the same site.

Another BWR with a passive cooling system (truly Gen-IV) is in the design and licensing phase in the United States. The U.S. Nuclear Regulatory Commission (NRC) has given this reactor's safety aspects a favorable review, which is a promising sign for American nuclear power.

Modular Small Reactors

Over the years many small reactors have been proposed and some have been built. They have been used to provide heat and power for remote army installations and in larger sizes to propel submarines and surface vessels of medium to large size (aircraft carriers). At present there is a renewed interest in these smaller systems. Both Toshiba and Babcock and Wilcox have discussed the provision of a small (ca. 30 MW) reactor for the city of Galena, Alaska. Babcock and Wilcox has recently announced its intention to obtain NRC certification for reactors placed underground. These would be of modular design with the individual units having power ratings of 125 MW. Spent fuel storage would also be underground and with refueling the reactor will have a presumed lifetime of 50 years.

Perhaps the most innovative of the new reactor concept is based on uranium hydride as a fuel. Conceived by Dr. Otis Peterson of Los Alamos National Laboratory, and being developed by Hyperion Power Generation, the technology is based on fission of uranium as uranium hydride, UH_3. With hydrogen as a chemical constituent of the fuel, this technology does not require external light water for moderating the neutrons. If the fuel gets too hot, UH_3 loses hydrogen, and the absence of a moderator reduces the fission cross section and the fuel cools down quickly, reabsorbing hydrogen. This self-regulating ability has the immediate benefit that this system does not require control rods and other moving parts to maintain its activity. The embodiments being tested consist of sealed reactors with radiation shielding that are entombed in concrete and buried underground. A circulating gas extracts the heat from the core, and an exchanger—external to the concrete structure—uses this heat to produce steam and generate power.

As the entire unit is a cylinder about 9 feet tall and a few feet across, it can be retrieved and shipped to a factory for refueling—needed every five years if it is operated at full power over the interval. The factory refueling proposal is novel, but it might create many questions as well as problems, similar to those now faced by utilities that would like to ship their spent fuel to a temporary or a permanent storage site,

Each UH_3-based 75 MW_t (thermal) or 25 MW_e (electric) unit is expected to cost about $25 million, and the estimated cost of electricity is between 5¢/kWh and 8¢/kWh. If successfully developed, the technology holds promise for quickly being able to provide clean power to sites remote from any grid such as for developing shale and oil sand resources or for powering many rural areas of the developing world. The exhaust heat could also be used to provide clean water. Many military installations that seek independence from the local power grid are likely early adopters of the Hyperion technology.

Global Nuclear Power Generation

Exhibit 6.1 lists the reactor types in commercial operation in 2006 along with their capacity and type of moderator and coolant. Some are fueled with natural uranium, but most require fuel slightly enriched in the ^{235}U isotope. The combined capacity of the 438 reactors in 2006 was 372.1 GW, and together they produced an estimated 0.16 CMO power at a world-average availability factor of 76%. The average availability factor of nuclear plants in United States in 2006–2007 was significantly higher, at about 90%.

As exhibit 6.1 shows, pressurized water reactors (PWRs) dominate the nuclear energy field with about 65% of the installed capacity; in fact, in 2006 they and boiling water reactors (BWRs) together accounted for 93% of the world's nuclear power generation capacity. The next largest contributors,

Exhibit 6.1.
Nuclear Reactors in Operation in 2006, by Type, Capacity, Fuel, Moderator, and Coolant

Type	Number	Capacity[a] (MW)	Fuel	Moderator	Coolant
PWR	263	241,139	UO_2, enriched	H_2O	H_2O
BWR	94	85,442	UO_2, enriched	H_2O	H_2O
PHWR/CANDU	45	23,991	UO_2, natural	D_2O	D_2O
LGR	16	11,404	UO_2, enriched	C, graphite	H_2O
AGR	14	8,380	UO_2, enriched	C, graphite	CO_2
GCR	4	1,414	U, natural	C, graphite	CO_2
LMFBR	1	560	UO_2, PuO_2	None	Na
PHWR	1	148	UO_2, natural	D_2O	H_2O
Totals	**438**	**372,478**			

[a]Capacity refers to net output to power grid, design rating. From *Nuclear News*, vol. *50*, no. 3, March 2007.

Abbreviations: C, carbon; D_2O, deuterium oxide; H_2O, water; Na, sodium, PuO_2, plutonium oxide; U, uranium; UO_2, uranium oxide; PWR—pressurized water reactor, light water moderated and cooled; BWR—boiling water reactor, light water moderated and cooled; PHWR/CANDU—pressurized heavy water reactor, heavy water moderated and cooled, based on Canadian design; LGR—graphite-moderated reactor, light water cooled; AGR—advanced gas-cooled reactor, graphite moderated, CO_2 cooled (advanced design); GCR—gas-cooled reactor, graphite moderated, CO_2 cooled (early design); LMFBR—liquid metal fast breeder reactor, no moderator, sodium cooled; PHWR—pressurized heavy water moderated, light water cooled reactor.

the CANDU-type heavy water reactors, contributed less than 4%.[11] In 2007, three new reactors came on line: one each in China (1000 MW, PWR), India (202 MW, PHWR), and Hungary (450 MW, PWR). These additions together with increases in the rated power for some U.S. reactors increased world nuclear capacity by 9%.

There is much talk about resurgence of nuclear power these days. The discussion is driven by the growing demand for electric power, and particularly for power free from carbon emissions. As mentioned above, the installed capacity of nuclear power grew at an average rate of 25% per year between 1960 and 1987, to about 350 GW. The installed capacity has essentially stalled at that level for the past 20 years but is now poised for rapid growth. The International Energy Agency (IEA) projects that electricity demand will double by 2030, and that would necessitate an additional 4,700 GW of installed capacity. With this demand, the World Nuclear Association (WNA), an international association of the nuclear power industry, has developed three scenarios. In their reference case, they project that installed nuclear capacity would increase to 524 GW by 2030.[12] In their high-growth scenario, the nuclear capacity could double to 740 GW, and at 85% availability this would produce about 0.3 CMO. It is important to note that even under the WNA's high-growth scenario, nuclear power will not be providing any greater share of electricity than it is currently providing, namely, about 16%. It will simply aid in the attempt to prevent electrical energy shortages. The reference and high-growth scenarios would entail construction of 200–400 new plants over the next 25 years, and this growth is predicated on nuclear power being both economical and acceptable to the public.

Projections of the future use of nuclear power are often related to its cost compared to that of coal-fired power. However, public opinion can have a marked effect on its selection. For example, Italy rejected nuclear power after installing two small and two larger nuclear units. This country closed its last plant following the Chernobyl accident. Italy now has 10 reactors in the "proposed" category. Belgium is highly dependant on nuclear electricity from 7 units but is not proposing to construct any more. Perhaps it will buy nuclear power from Germany or, more likely, from France, if Belgium's expectations of producing larger amounts of renewable energy and achieving greater energy conservation are not fully realized. Great Britain has been slowly closing its earlier nuclear units but is now planning to build four newer, much larger ones to retain a sizable nuclear power production. Sweden has debated the use of nuclear power for several decades—its first referendum on the nuclear option called for elimination of its twelve plants by 2010. Sweden currently

---

11. We do not have data on power production by reactor type, but the fraction of nuclear energy produced by the leading types is probably equal to or greater than the capacity ratios.

12. World Nuclear Association Working Group, R. W. Gale, Chair. *The New Economics of Nuclear Power,* World Nuclear Association, London, 2007.

generates 42% of its electricity from the ten nuclear units still operating, while seemingly still debating their use. In the first half of 2009 the tide has turned in the favor of nuclear power, but at the moment the decision to build new units, or upgrade older ones is under discussion in Sweden. Japan, despite a wholehearted devotion to development and utilization of solar technologies, is now building two nuclear units and plans to build 15 more with a total capacity of over 20 GW. Japan now obtains one-quarter of its electricity from nuclear units.

The emerging Asian nations, China and India, will be the most aggressive in pursuing nuclear electricity. Both are now heavily dependent on coal-fired plants for their power. While currently producing about 2% of their electricity from nuclear installations, they both have plans to substantially increase the nuclear contribution. In mid 2009, China had 15 units with a capacity of 15.4 GW under construction, and was planning to build 34 more with a capacity of 36.4 GW. When completed, China would have 62 reactors with a combined capacity of over 95 GW, almost as large as the 101 GW installed capacity in the United States.

Beyond economics of power generation and public acceptance, a shortage of construction materials and of skilled engineering and construction services would limit the rate of expansion of nuclear power. We examine the factors affecting the cost of nuclear power in the next section, and later in this chapter we discuss the concerns about nuclear power that have hindered public acceptance.

Nuclear Power Economics

Nuclear power costs are dominated by the high capital cost of constructing the generating units—a function of almost all countries' requirements that reactors have containment structures (massive structures surrounding the reactor building to contain radioactivity within the plant in the event of an accident) and redundant safety systems. As noted above, nuclear fuel costs are relatively low, around 15% of the cost of power, in contrast to those used for fossil-fired electricity generation (typically around 30% and 70% for coal and natural gas respectively) but higher than those for income sources of electricity whose "fuel costs" are zero.

Because power plants can take many years for construction, the cost of money, or the interest on the investment over the period, becomes a significant cost factor. A principal reason for the long construction time has been the bottlenecks in the manufacture of critical components, as well as in the design, engineering, and construction of critical plant facilities. In the United States, many of the plants have been one-of-a-kind, and therefore the approval processes have also been long. We can expect that if the designs were standardized, limitations from both manufacturing bottlenecks and regulatory approval would be eased.

The construction costs are often quoted as instant costs, as if the entire construction took place overnight. The instant construction costs for nuclear power units are high, ranging from $1,000/kW to $5,000/kW, depending on the specifics of the proposed system. We must add to these instant costs the interest that must be paid on the investments. Thus, projected and actual costs of nuclear power depend not only on the basic instant construction estimates but also on the prevailing interest rates over the construction period and on the times between the initial commitment to the purchase of long-lead time equipment, the actual start of construction, and the time before reliable full-power operation is achieved.

Exhibit 6.2 illustrates the effect of construction time and interest (or discount) rates on the effective capital costs of any power installation. Gas-fired power plants require an average construction time of two years, coal-fired units from two to four years, and nuclear units from three to five or more years. Many nuclear power plant projects in the United States have experienced long delays. Some delay was caused because lack of standardization has meant developing many one-of-a-kind components, and others because the approval process—exacerbated by the lack of a standard design—had to be more thorough and time-consuming. The delays have resulted in cost overruns.

The WNA report also summarizes estimates made by six independent groups from different countries on the economics of power generation, comparing the costs of electricity from nuclear with that from coal and natural gas. All of these estimates were conducted around the same time (2003–2005), but the different groups used different discount rates as well as varying requirements for the construction, reflecting their local realities. The capital costs ranged from a low of $1,058/kW to a high of $2,082/kW. In the future, capital costs may be lower if the designs are standardized so that the approval can be streamlined. GE and Westinghouse have submitted new reactor designs for NRC review that are both simpler and more robust than existing ones. The elimination of excess piping, valves, and controls in these designs is expected to reduce field construction costs. Moreover, these two companies as well as others are investigating the use of factory-constructed modules that can be

Exhibit 6.2.
Effect of Construction Time and Interest Rate on the Capital Cost of Nuclear Power (percent above instant construction costs)

Years of Construction	Discount Rate	
	5%	10%
3	10%	21%
5	16%	34%
7	20%	49%

Exhibit 6.3.
Costs to Prepare One Pound of Reactor-Ready Uranium Fuel (data rounded)

	U.S. Dollars per Pound	Percentage of Total Cost
Ore Purchase	$214	26
Conversion to UF_6	41	5
Enrichment	448	55
Production of Fuel Pellets	109	13
Preparation Total	$812	99

made more cheaply and put into place in the field with less time and effort than is currently the case. As noted above, three- to four-year field construction times have been reported for the construction of ABWRs in Japan.

On the other hand, we should also note that in the last four years, construction costs have increased dramatically in response to a rapid rise in the cost of many key inputs such as steel and cement that have appreciably outpaced general inflation. Steel prices rose almost 10-fold between 2006 and 2008. This general increase will affect not only nuclear power but also competing coal- and gas-fired power plants as well as other power generation systems such as wind and solar.

Fuel and operating costs are generally low for nuclear installations. A 2007 WNA study traced the cost for preparing a pound of reactor fuel, from procuring the ore through milling, converting to UF_6, enriching the isotope, and preparing the fuel pellets. The results are shown in exhibit 6.3. It generally requires ore containing seven pounds of uranium to produce one pound of enriched uranium fuel. Taking into consideration other losses along the steps, the study found the total cost to produce a pound of fuel was $812.[13] In a typical PWR this pound of fuel would produce 155 MWh of electric power, and thus the cost of fuel works out to about 0.5¢/kWh, with the original uranium ore accounting for only 26% of that cost. The enrichment process to produce finished, reactor-ready fuel assemblies from raw ore is the most expensive fuel cost component, representing about 55% of the cost of reactor-ready fuel. These fuel costs were arrived at assuming a uranium ore price of $53 per kilogram of U_3O_8 ($24/lb U_3O_8), which was about one-third of the spot price— the price paid for immediate delivery and payment as opposed to contract prices that are negotiated for future delivery and transfer of funds. Even if the ore were purchased at twice the assumed price—two-thirds of the spot price—it would increase the total price of fuel to only $1,039 per pound or 0.67¢/kWh.

13. World Nuclear Association. *The Economics of Nuclear Power,* August 2008. The study reports the cost for a kilogram of the fuel that we have converted to cost per pound.

Exhibit 6.4.
Comparative Cost Breakdown for Electricity Production in Finland (euro/MWh, 2003)

Cost Factor	Nuclear	Natural Gas	Coal	Peat	Wood	Wind
Capital	13.8	5.3	7.6	10.2	13.0	40.1
Operation and Maintenance	7.2	3.5	7.4	6.5	8.2	10.0
Fuel	2.7	23.4	13.1	17.9	23.1	0.0
Total	23.7	32.2	28.1	34.6	44.3	50.1
CO_2 Penalty (20 Euro/te)	0.0	7.0	16.2	19.6	0.0	0.0
Total with CO_2 Penalty	23.7	39.2	44.3	54.2	44.3	50.1

As the above example illustrates, on average uranium cost comprises around 15% of the cost of electricity from nuclear plants. Contrast that with a natural-gas-fired plant, where fuel costs amount to 75% of the cost of electric power. Whereas doubling the price of natural gas increases the cost of electricity from a gas-fired plant by 70%, a similar doubling of uranium price increases the cost of nuclear power by only 5–10%. Continued improvements in the efficiency of power generation and the extent of fuel burn-up through better management of the process should lead to even lower fuel costs for nuclear plants.

In the six studies summarized in the 2007 WNA report mentioned above,[12] the range of projected cost of electricity was from 2.7¢/kWh to 6.7¢/kWh.[14] These estimates of the cost of electricity include the payments nuclear plant operators have to make for future handling of nuclear waste as well as maintaining liability insurance. A 2003 Finnish study, which was included in the WNA report, provided the breakdown of capital, operation and maintenance, and fuel costs for producing electricity in Finland by nuclear, coal, gas, peat, wood, and wind. Their results are displayed in exhibit 6.4.[15] The values for the cost of electricity take into account the different availability factors for the different sources of power, and therefore are a more realistic metric of comparison than simply the capital cost per megawatt of installed capacity. Capital costs for nuclear power plants were more than for natural gas, coal, and peat but were considerably less than for wind. The higher capital cost was more than compensated for by the lower fuel price, with the result that nuclear power was less expensive than power from any other method, and even more so if a penalty of 20 euro/tonne of CO_2 was levied for the emissions from coal, natural gas, and peat.

14. The studies cited costs in local currencies, and we have converted them using the average exchange rate for the year of the study.

15. R. Tarjanne and K. Luostarinen. *Competitiveness Comparison of the Electricity Production Alternatives*, Research Report EN B-156, Lappeenranta University of Technology, 2003.

Exhibit 6.5.
Projected Costs of Electricity Production Using Coal and Nuclear Fuels
for Selected OECD Countries (¢/kWh, 2003; 40-year plant operating life
and 85% capacity factor assumed in all cases)

| | Discount Rate | | | |
| | 5% | | 10% | |
Region and Country	Coal	Nuclear	Coal	Nuclear
North America				
Canada	3.51	2.60	4.12	3.71
United States	2.71	3.01	3.65	4.65
Europe				
Czech Republic	2.94	2.30	3.71	3.17
Finland	3.64	2.76	4.45	4.22
France	3.33	2.54	4.43	3.93
Germany	3.52	2.86	4.09	4.21
Netherlands	—	3.58	—	5.32
Romania	4.53	3.06	5.15	4.93
Slovakia	4.78	3.13	5.52	4.55
Switzerland	—	2.88	—	4.38
Asia/Pacific				
Japan	4.95	4.80	6.91	6.86
South Korea	2.16	2.34	2.71	3.38

A recent joint study of the IEA and the Organisation for Economic
Co-operation and Development's (OECD) Nuclear Energy Agency (NEA) also
estimated costs for coal- and nuclear-generated electricity and found the costs
for nuclear power to be lower over time.[16] As shown in exhibit 6.5, 12 countries were studied with costs accumulated over a five-year construction time
and with discount rates of 5%/yr and 10%/yr. At the 5% discount rate, nuclear
power was less expensive than coal in 10 of the 12 countries studied, the excep-
tions being the United States and South Korea. At 10% discount rate, nuclear
power was still competitive in 9 of the remaining 10, with Germany joining
the United States and South Korea as countries where electricity from coal
was cheaper. This study estimated instant nuclear plant capital costs as low
as $1,000/kW (in the Czech Republic) and as high as $3,500/kWh (in Japan).
The average instant capital cost assumed was $1,500/kW.

The electric power generated in *existing* nuclear units is currently cheap,
given that the capital costs of many operating units have been fully depreciated.

16. Nuclear Energy Agency. *Projected Costs of Generating Electricity, 2005 Update,* Organisation for Economic Co-operation and Development, 2005. http://www.nea.fr/html/general/
press/2005/2005–02.html. Accessed August 2007.

The additional costs required to extend a reactor's licensed operating span another 20 years in the United States will be only a fraction of the plant's original costs. Even with inflation, these additional costs will be fully depreciated before the reactor reaches the end of its useful life—as many as 50–60 years after first license approval. Moreover, if a carbon tax were imposed on the CO_2 formed during coal combustion in coal-fired plants to produce electricity, with CO_2 captured at the plant stack and stored in caverns or in the ocean, the electricity produced would also be more expensive than that from nuclear power plants. Penalties for CO_2 emissions of 20–30% over the base generation cost have been suggested.

Greenhouse Gas Emissions from Nuclear Power

Nuclear power is generally considered to be free from greenhouse gas emissions. There are some emissions associated with the construction and decommissioning of the plant, as well as with the production of the fuel from the ore. Energy from a variety sources, including coal and diesel, is often used in such operations. A life-cycle analysis of power generation from fission, fusion, coal, and wind power plants published in 2000 compared the projected energy payback periods and the associated CO_2 emissions, and found that CO_2 emissions from fission plants were the same as from wind power at 15 g/kWh.[17] However, a report by Jan Willem Storm van Leeuwen, secretary of the Dutch Association of the Club of Rome,[18] projects that total greenhouse gas emissions from these processes amount to more than 100 g CO_2/kWh. This value is far from the near zero CO_2 emissions that most media reports would suggest, although even this high estimate of CO_2 emissions is only a tenth of that from coal-fired power plants. This report has been widely cited by others to bolster opposition to nuclear power, and it behooves us to examine the assumptions that led to the surprisingly high value of CO_2 emissions.

The basic premise of van Leeuwen's calculation of CO_2 emissions is that over time uranium-rich resources (greater than 0.1% uranium oxide) will be exhausted, and use of ores that are deficient in uranium (between 0.001% and 0.002% uranium oxide) will require substantially larger amounts of energy inputs for preparation of fuel. About 40 g CO_2/kWh of emissions were associated with the production of the fuel itself in his calculations, and this number is likely to grow if ores with lower concentration of uranium are used. In his

17. S. W. White and G. L. Kulcinski. "Birth to death analysis of the energy payback ratio and CO_2 gas emission rates from coal, fission, wind, and DT-fusion electrical power plants," *Fusion Engineering and Design*, vol. 48, pp. 473–481, 2000.

18. Jan Willem Storm van Leeuwen. "CO_2 emissions from nuclear power," in Frank Barnaby and James Kemp, eds., *Secure Energy? Civil Nuclear Power, Security and Global Warming*, Oxford Research Group, London, 2007, pp.40–44.

analysis, he considers the ore requirements based on a simple once-through cycle and does not consider the potential offered by breeder reactors, because these technologies are not fully commercialized. The energy inputs for these latter systems are 1/40th to 1/50th (2–2.5%) of the energy inputs of the once-through cycle. These options are currently not seriously considered because uranium inventories are high and the ore is still plentiful and cheap.

The main source of the large CO_2 emissions in van Leeuwen's analysis is in the front-end mining of uranium ore, enriching it, and preparing the fuel, amounting to 40 g/kWh. Critics of his conclusions are quick to point out that actual energy costs in mining are much lower than those he assumes, and therefore the CO_2 emissions are also much lower. He has further neglected to take into account the construction material requirements for new reactor systems. These are somewhat less than those of current systems.

Under more appropriate assumptions of longer reactor life and lower energy requirements for mining, as well as the use of centrifuges instead of diffusion enrichment (a gain of efficiency of approximately fivefold in that step), modern reactors with their longer operating life will have much lower CO_2 emissions per unit output than assumed by van Leeuwen.

Instead of using more theoretical calculations, we can turn to actual data from Vattenfalls, a Nordic company that operates nuclear, hydro, coal, wind, and gas-fired power generation plants. In accordance with E.U. regulations, Vattenfalls has filed an environmental product declaration, an independently audited inventory and assessment of CO_2 emissions associated with the construction, operation, dismantling as well as of front-end fuel production and back-end storage based on the experience of three BWR plants at Forsmark, Sweden. The three plants have a combined capacity of 3 GW. According to this declaration, the CO_2 emissions are only 3.67 g/kWh, which is even less than the estimate of 15 g/kWh that the 2000 report gave for fission plants and wind turbines.[17] Vattenfalls also reports that on a life-cycle basis, wind power produces 13.5 g/kWh of CO_2, about 75% of which results during the construction of the turbine. We normally think of nuclear plants as requiring large quantities of cement and steel to provide the containment structure, and therefore it might come as a surprise that CO_2 emissions from construction of a nuclear plant are less than those for a wind generator. We find corroboratory evidence for low CO_2 emissions during construction of nuclear plants in other reports.[19] The relatively low emissions reflect the fact that while a single nuclear facility can generate 1 GW power and produces electricity 85% of the time, it takes more than 200 of the largest wind turbines—5 MW—to equal that capacity, and they produce power at best for about 30% of the time. Thus,

19. See, for example, World Nuclear Association, *Energy Analysis of Power Systems*, 2006; and Joop F. de van Vate, "Full-energy-chain greenhouse-gas emissions: A comparison between nuclear power, hydropower, solar power and wind power," *International Journal of Risk Assessment and Management*, vol. 3, pp. 59–74, 2002.

greenhouse gas emissions from nuclear power are essentially zero, being even less than those from wind power.

The Resource Base: Uranium, Thorium, and Deuterium

Uranium

As with fossil resources, the uranium resources are also estimated with varying degrees of confidence and at varying prices. Several agencies have historically published the quantities of uranium ores, but they have used different price points and a somewhat differing system of classifications, making direct comparisons difficult. In 2003, the NEA published its assessments based on the data from 56 reporting countries.[20] The amount of uranium in the highest confidence category, "reasonably assured resources" (RAR), available at an acceptable cost (i.e., less than $130/kg or $59/lb) was estimated to be 3.5 million tons (10.6 CMO). The NEA also considered it possible to produce an additional 1.5 million tons in the category with the next confidence level also at the same price point of less than $59/lb. The sum total of these two top classifications or "known conventional resources" is 5.0 million tons-U, roughly equivalent to 15 CMO if the uranium is used in the common once-through LWR reactors.

The distribution of the relatively more easily obtainable 3.5 million tons in the RAR category of uranium resources by region is shown in exhibit 6.6. By country, these resources are located, in descending order of quantity, in Kazakhstan, Australia, Canada, South Africa, and the United States. Of the other countries that have useful uranium deposits, only Algeria, with 21,450 tons, is in the Middle East/North Africa region, which dominates our oil and gas resources. Estimated additional resources add some 1.5 million tons to the total of easily obtainable uranium, and larger amounts are potentially available at higher costs. The availability of additional resources will likely be increased with increasing intensity of exploration, but the large amounts of known conventional resources do not currently warrant intensified exploration. It has been estimated that doubling the price of uranium ore, which as mentioned above would increase the cost the electricity by less than 1¢/kWh, would increase the known conventional resources base 10-fold.[21] This increase could easily support a much larger (CMO/yr-level) role of nuclear power for several centuries.

The resource estimates shown in exhibit 6.6 are based on the data reported in the OECD *Red Book* cited above, which provides resource estimates from

20. Nuclear Energy Agency. *Forty Years of Uranium Resources, Production and Demand in Perspective: The Red Book Retrospective,* Organisation for Economic Co-operation and Development, Paris, 2002.

21. John Deutsch and Ernest Moniz, cochairs. *The Future of Nuclear Power: An Interdisciplinary MIT Study,* Massachusetts Institute of Technology, Cambridge, Mass., 2003.

Exhibit 6.6.
Reasonably Assured Uranium Resources at $59/lb-U or Less

Region	Thousand Tons	CMO	Percentage of World Total
North America	770	2.3	21%
Central/South America	105	0.3	3%
Europe	50	0.2	2%
Russian Group	900	2.7	26%
Asia/Pacific	956	2.9	27%
Middle East/North Africa	22	0.1	1%
Sub-Saharan Africa	684	2.1	20%
World Total	3,486	10.6	100.0%

the participating 109 countries. Because Greenland is a part of the North American continent, we have included its roughly 0.06 CMO of nuclear resources with North America, and in that regard we depart from other compilations that include them with Europe. As we discussed, the costs of the uranium ore resource constitute only a small fraction of the overall cost of nuclear power, in contrast with fossil fuel electricity production.

The energy obtainable from a given amount of uranium depends on the design of the reactor and the uranium preprocessing. Most of the reactors use a pressurized water configuration and use slightly enriched uranium. They require 328,000 tons of natural uranium per CMO of energy produced. At current rates of use (0.15 CMO/yr), the assured uranium resource of 3.5 million tons would last more than 70 years. This assumed longevity of supply is deceptive, however. On one hand, the small renaissance in nuclear power production occurring now is expected to increase fuel consumption in future years and thus reduce the longevity, and on the other hand, with increasing uranium costs there is prospectively much more uranium.

Three potential additions can be made to these relatively assured *in-earth* resources. First, there is the uranium already mined and used in power production. The major nuclear reactors that use enriched uranium discharge spent fuel containing substantial quantities of uranium with a higher ^{235}U content than natural uranium ore (about 0.9% instead of the natural 0.71%). This spent fuel can be processed and the uranium (and associated plutonium) recovered and used. A second additional source of uranium is the weapon stores of several nations, which contain both uranium and plutonium. This resource is relatively small, with a fraction of a CMO in the weapon stockpiles of the seven declared nuclear nations.[22] The Russian Federation and the

---

22. *Red Book* data indicate that about 2.4 million tons of uranium had gone either to nuclear weapon producers or to civilian inventories by 2002; perhaps 90% of that production was for civilian power purposes.

United States, which have the largest stockpiles, are slowly releasing a portion of this stockpiled material for peaceful use.

The third additional source is not a source *per se* but improved methods for nuclear power production that can increase the usefulness of the uranium stock. This addition is our largest resource because the power produced by a single unit of uranium fuel can be approximately doubled by recycling the plutonium produced in the first cycle combined with the residual uranium in the spent fuel in a second load fed to a reactor—a process that could be repeated several times by making proper adjustments to the reactor fueling and operations cycles. Further, if breeder reactors were used, the energy extracted from the initial uranium mined could be 40–50 times that produced by simple once-through use in conventional reactors.[23]

At its $59/lb spot market price in 2007, uranium as a nuclear fuel was competitive with oil at about $20/bbl or coal at $25/t.[24] The price of coal, more than of oil, is the determining economic factor here because both coal- and uranium-fired units produce significant amounts of base-load electric power. The excursions to more than $100/bbl for oil in the spot market and perhaps $50/ton for coal make the higher uranium prices more competitive. Moreover, the development of the new centrifuge enrichment technology has reduced the cost of enriching fuel substantially, and the cost of any uranium fuel in current use is a small fraction of the overall cost of nuclear electricity. These factors alone enable the use of higher cost, higher quality ore. In addition, recycling of spent fuel at an additional cost could nearly double the usefulness of a given quantity of ore, the use of normally spent or natural uranium fuel in breeder reactors could raise fuel availability by 40–50 times, thereby lowering effective fuel prices substantially. Thus, at a nominal rate of increase in nuclear power use, say, 5%/yr, the current assured uranium supplies could be extended to perhaps 100 years and still be available at reasonable cost.[25] In addition, the use of breeder reactors would allow thorium to be converted to fissile ^{233}U, further increasing overall nuclear fuel availability, because thorium could substitute for uranium in the blanket region and fissile ^{233}U would be produced.

Thorium

Thorium is 3.5 times as common as uranium in Earth's crust. There are large thorium deposits within the North America, Asia/Pacific (especially India and

---

23. Minor adjustments would be needed in the level of enrichment used in second-load fuel to account for the presence of the other uranium isotope, ^{236}U.

24. Note that most of the uranium is traded with long-term contracts at about half of the spot price. This situation is not unlike that of oil, which is also traded at contract prices significantly different than the spot price quoted in the media.

25. Authors' estimates.

Australia), South America, and Middle East/North Africa regions. Turkey, a part of Europe in our groupings, also has significant deposits. The stable and sole isotope of naturally occurring thorium is ^{232}Th. It can absorb neutrons to produce ^{233}U, a fissile isotope with reasonably long half life. Fission of ^{233}U can then be used for nuclear power production; in the nuclear jargon, this makes thorium a "fertile" material like ^{238}U, which can be bred into fissile ^{239}Pu. Fuel production for processes using the thorium cycle does not require isotope enrichment. There has been only limited experience with the use of thorium and the breeder technology for utilizing it is not mature.

If commercialized, an advantage of using thorium would be that it produces far less transuranic waste (Pu, U, and other materials) that could be processed into nuclear weapons than does uranium. Deployment of the thorium cycle would perhaps decrease the possibility of undesired nuclear proliferation. A potential downside to its utilization includes the need for substantial shielding—and the attendant high cost—during fabrication of ^{233}U fuel from irradiated thorium that contains highly radioactive ^{228}Th, an alpha emitter with a relatively short half life of 2 years.

Deuterium

As we have stated, deuterium—the heavy hydrogen isotope—is abundant in nature. It is naturally present in water at 0.015% of all hydrogen atoms, and thus may be the long-term solution for the generation of nuclear power through fusion. If all of the deuterium contained in the oceans could be extracted and fused in what is known as a deuterium-deuterium (D-D) reaction, the energy theoretically available would amount to many trillion CMO. However, D-D reactions are more difficult to achieve than other fusion reactions, and scientific attention centers on the use of deuterium-tritium (D-T) fusion schemes because they are potentially easier to implement. The D-T fusion is more facile, but tritium is not a naturally abundant isotope of hydrogen. It has a relatively short half-life of about 12.3 years. For the current experimental D-T fusion reactors, tritium is bred from lithium (^{7}Li) by reaction of ^{7}Li with a neutron. Thus, according to the current assessments, it is lithium availability that could limit the ultimate amount of fusion energy.[26]

Concerns about Nuclear Power

Many people fear nuclear power because of the word "nuclear" and what it connotes. The U.S. use of nuclear weapons on Hiroshima and Nagasaki that

---

26. The amount of lithium in seawater is vast. Although theoretically it may be sufficient for many billion CMO, the practicality of such extractions will have to be considered.

ended World War II resulted in major loss of life—both immediately and for years thereafter. The effects of the two crude and low-yield bombs used left gruesome images of unprecedented levels of destruction—apparent almost instantaneously—and a lasting, unfavorable impression in the public mind. It is important to note that the deaths and destruction were mostly a result of the blast, not the radioactivity that persisted. People close enough to receive the lethal radiation could not have survived the blast and heat. The three-day bombings of Dresden by allied forces also left tens of thousands dead, and the numerous fire bombings of Tokyo and other Japanese cities by the U.S. Air Force resulted in large but never officially counted deaths—believed to number in the hundreds of thousands at least—but did not have the same visual coverage as the two nuclear blasts.

Relatively small accidents at experimental and early facilities for making plutonium or for power production created minor hazards to the public that went largely unnoticed. The widely publicized accident at Three Mile Island, Pennsylvania, did essentially no harm to the public, and although the reactor itself was destroyed, the containment vessel was not. The accident at Chernobyl, in the Soviet Union, involved a reactor without containment protection and that, in hindsight, had design flaws. It was also being used in an experimental procedure conducted by technicians without apparent knowledge of the basic science underlying the reactor's performance and capabilities. Here, the primary accident was not a nuclear one but a steam explosion that occurred when water contacted hot metal, producing hydrogen, which in turn reacted with oxygen in a second explosion. These explosions ejected radioactive material into the atmosphere, which wreaked terrible damage upon the surrounding area. Fortunately, with modern safety measures, a similar incident is considered essentially impossible today.

Nuclear hazards have received much attention in the media, mostly unfavorable to the use of nuclear power. On the other hand, this power source offers a potentially long-serving energy source, and one particularly well suited to fill a considerable part of the world's continually increasing demand for electricity. We therefore believe the concerns expressed about its hazards need to be addressed in any discussion about its suitability. These hazards fall under three categories: (1) fears of radiation and explosions associated with the operation of nuclear reactors themselves; (2) political concerns, which refer to the risks of nuclear proliferation and terrorism and require a political solution; and (3) technical issues, which involve the long-term storage and safeguarding of nuclear waste.

Radiation

Exposure to radiation, specifically ionizing radiation, is known to cause harmful radiation illness. It can also cause cancer, even when acute toxic effects are not evident. Ionizing radiation consists of highly energetic electromagnetic waves, such as X-rays and gamma rays, as well as particles that can ionize molecules (remove electrons) as they move through a medium. Because many of the materials involved in nuclear power production emit ionizing radiation, many people view nuclear power negatively. Yet, we are exposed to naturally occurring ionizing radiation on a daily basis. In contrast, properly operating nuclear power plants do not emit radiation beyond the plant boundaries, and living next to one does not increase our exposure to radiation. You cannot locate a nuclear power plant with a Geiger counter, because the normally occurring background radiation is too large. Leaks are a different matter, particularly in the context of long-term storage of radioactive waste. We discuss that problem a little later in this section.

Although fears about radiation are sometimes justified, they should be put into the context of our everyday lives. Our most common source of radiation exposure is carbon. Cosmic rays interacting with nitrogen in the upper atmosphere produce a radioactive isotope of carbon, ^{14}C. This radiocarbon is present in the atmospheric CO_2 to the level of one part per trillion, which plants absorb and animals acquire through feeding on the plants. Other sources that bathe us daily in a "sea" of radiation include:

Cosmic radiation from space

Radiation associated with naturally occurring uranium and thorium in rocks, including granites and shales, which we use as building materials

Radiation from radon gas (including its isotope thoron) that is produced in the rocks bearing uranium and thorium

Occupational exposures (e.g., for miners who work underground or for people who fly frequently)

X-rays during medical and dental examinations and treatments

As mentioned earlier, decay of radioactive isotopes leads to emission of alpha or beta particles, as well as high-energy X-rays or gamma rays. These emissions differ in their ability to generate ions in any given medium. It depends largely on their energy, how penetrating they are, how much energy they deposit, and what tissues in our bodies are exposed. Radioactivity can be measured in terms of the number of decays per second (curies), or the amount of ionization the radioactivity can cause (roentgens) in a given volume of a reference medium—air. However, when discussing the potential damage from ionizing radiations, it is useful to take into account variations in energies of different radiations, their penetration depths, and the response of different organs to the radiation. While many organs in our bodies can tolerate loss of a considerable number of cells, at some point the effects of harmful levels of exposure become evident. Biologists express the effective absorbed dose in the units of rem (roentgen equivalent in man).[27]

Two types of exposures concern us here: (1) acute exposure to relatively large doses and (2) cumulative effects of low levels of exposure. An acute

---

27. Another unit for radiation exposure often used by physicists is the sievert ($1 \, Sv = 100 \, rem$), but most media discussions of radiation exposure use rem.

exposure of 100 rem is barely noticeable by individuals. Exposures of about 200 rem are likely to cause nausea, radiation sickness, and loss of hair. These are symptoms that people undergoing radiation therapy often display. For ionizing radiation, an acute exposure of 300 rem is the LD_{50}, which refers to a 50% likelihood that this will be a lethal dose, within two months, for a person who was unfortunate enough to have received that level of exposure. Acute exposure to more than 1,000 rem is generally lethal within a few days.

One rem of radiation exposure would correspond to a body receiving about 10 trillion average-energy gamma rays. Nevertheless, the total amount of energy deposited by one rem of radiation is miniscule, and if it were to dissipate into heat, our body temperature would increase infinitesimally, perhaps a billionth of a degree. Now, if the ionizing radiation were to deposit its energy into a single molecule, it could easily break bonds. Further, if that molecule happened to be a gene that regulates the cell division process, then disrupting such molecules may increase our chances of contracting cancer. The current consensus among medical professionals is that an exposure of 25 rem increases the risk of cancer by 1%. This number is arrived at from studies with considerably larger doses than 25 rem and extrapolating the results to low levels assuming a linear response. As we just stated, 25 rem is an amount that our bodies will hardly notice, and most likely any damage done will be healed by the body's repair mechanisms. However, there are multiple genes that determine the onset of cancer, so it is possible that for someone who is at the verge of getting cancer, even a single additional damage may trigger the onset. A 1% increase in cancer from an exposure of 25 rem means that if 100 people were to receive that dose there would be one *additional* case of cancer among them. We must emphasize additional, because cancer is not a rare disease; it affects about 20% of the population, and therefore in a population of 100 people, we would as such expect about 20 cases of cancer from all other reasons.

With this background information on the toxicity of radiation, let us now look at the amount of ionizing radiation we are exposed to in our daily lives, as a means of gaining perspective to place the exposure from nuclear plant operations and accidents. Note that while we have so far been discussing doses of several rem, the following discussions on ambient exposures are in thousandths of rem (millirem, or mrem).

Annual global average exposure from cosmic radiation is 40 mrem, and it ranges between 30 and 100 mrem. It is generally higher near the poles, although some variation is also observed longitudinally. Altitude also plays a role, because the air above us shields us from the radiation to some extent. At sea level, the annual dose from cosmic radiation in most of the United States is 26 mrem, increasing to about 47 mrem at an altitude of 5,000 ft (~1 mile).[28]

28. The American Nuclear Society has an interactive Web page to determine the annual radiation dosage at various locations in the United States (see http://www.ans.org/pi/resources/dosechart/). Accessed November 2008.

In Denver, the mile-high city, there is the additional radiation amounting to about 63 mrem from the uranium-bearing Colorado plateau and the radon that seeps out of it, bringing the total to 110 mrem. Apart from exposure to these external sources, we also ingest radiation. As stated at the start of this section, there is radioactive CO_2 all around us that is fixed by the plants and eaten by animals and us. We receive about 40 mrem of radiation from our food (radioactive carbon and potassium). Does it mean we are radioactive? Yes, indeed, albeit only slightly. The radon from the Colorado plateau mentioned above is *additional* radon that someone living there might get exposed to. But radon as a constituent of air is present everywhere, and we receive about 200 mrem of radiation from it annually. Thus, annual exposures of 300–400 mrem are a part of living on Earth. Living within 50 miles of a nuclear plant increases the annual dose by 0.01 mrem, and living within 50 miles of coal-fired power plant adds 0.03 mrem. Thus, living near a coal-fired power plant exposes us to more radiation than living near a nuclear plant, and in either case the increase is less than about 0.01% of our natural exposure.

Our exposure to radiation is also dependent on our occupation and other activities. Someone who works 40 hr/week for 50 weeks/yr in the concourse of New York's Grand Central Station receives 180 mrem while at work. Frequent fliers and airline pilots and cabin crew receive considerably greater annual dose of radiation, estimated at 300 mrem per person annually, in addition to the normal exposure from living on Earth. A single five-hour flight from San Francisco to Washington, D.C. exposes passengers to 3 mrem. A single X-ray at a dentist's office may expose us to 1 mrem, and a whole-body CAT scan exposes us to 100 mrem. Average yearly exposure in the United States from all natural sources is estimated at 300 mrem, while the average exposure from all man-made sources is 70 mrem. The largest dose of radiation that anyone offsite received from the accident at Three Mile Island has been estimated at 5 mrem. Using 300 mrem annual exposure as an example, and assuming a linear response, we see that over a lifetime of 80 years, such exposure would mean a cumulative exposure of 24,000 mrem (or 24 rem), which would increase the risk of cancer by about 1 percentage point, from 20% to 21%. Higher elevation leads to higher exposure, yet it is also the case that the cancer risk in Denver is lower than in the rest of the continental United States. This observation and other anomalies have prompted some to suggest that a small amount of radiation exposure may be beneficial, although no definitive studies have supported such a "hormesis" effect. Residents of some parts of Sweden receive a significantly higher exposure, about 3,500 mrem, and residents in parts of Tamil Nadu, India, where thorium-rich monazite sand is plentiful, receive an annual exposure of 5,300 mrem.

In a 2000 report, the United Nations Scientific Committee on the Effects of Atomic Radiation (UNSCEAR) summarized per capita radiation exposures associated with the numbers of persons involved in various activities,

Exhibit 6.7.
Summary of Occupational Radiation Exposures

Source/Practice	Number of Workers Exposed (thousands)	Average Annual Dose per Person (mrem)	Additional Cancers[a]
Man-Made Sources			
Nuclear fuel cycle	800	180	5.8
Industrial uses	700	50	1.4
Defense	420	20	.34
Medical uses	2,320	30	2.8
Education/veterinary	360	10	.14
Total from man-made sources	**4,600**	**60**	**11**
Natural Sources			
Flight crews	250	300	3.0
Miners (other than coal)	760	270	8.2
Coal miners	3,910	70	11
Mineral processors	300	100	1.2
Aboveground workers[b]	1,250	480	24
Total from natural sources	**6,500**	**180**	**47**

[a] Assuming linear response.

[b] Exposure of workers to radon.

as shown in exhibits 6.7 and 6.8.[29] These activity-based exposures can be compared to UNSCEAR estimates of radiation exposure of the general world population of 6.5 billion people. The natural exposure varies between 100 and 1,000 mrem with an average of 240 mrem. To calculate the number of annual cancer cases attributable to natural background radiation, we first multiply the world's population by 240 mrem to arrive at the total human exposure of 1.56 billion person-rem. We then we divide this total exposure by 2,500 rem, because according to the linear hypothesis, exposure to 25 rem increases the chances of cancer by 1%. That arithmetic tells us that across the globe 624,000 cases of cancer are attributable to background radiation annually. As shown by the UNSCEAR data, nuclear fuel operations expose 800,000 workers to radiation of 180 mrem and result in roughly 60 additional cancer cases. It is significant to note that radiation exposure incurred by nuclear fuel operators is less than that for workers in the general mining and airline industries.

29. United Nations Scientific Committee on the Effects of Atomic Radiation. *UNSCEAR 2000 Report to the General Assembly,* United Nations, New York, 2000.

Exhibit 6.8.
Exposure of the public to Radiation

Source	Average Exposure per Capita (mrem)
Natural Background	240
Diagnostic Medical Examinations	40
Atmospheric Testing of Nuclear Weapons	0.5
The Chernobyl Accident	0.2
Nuclear Power Production	0.02
Total	**281**

The most serious nuclear accident was from an explosion at the Chernobyl nuclear power plant in Pripyat, Ukraine, in the early morning hours of April 26, 1986. This plant had a graphite-moderated light-water-cooled reactor. These systems require that large quantities of water be continually circulated through the core, even when the system is not operating. A safety test was planned for the fateful day with the objective of testing out a scheme for safe operations in the event of an external power failure that would require the reactor to be shut down. There were diesel-powered systems that could generate the required power, but they took about a minute to come up to speed. The specific question this experiment was designed to answer was whether the steam turbines could provide stop-gap power for the water pumps during their spin-down cycle, to cover for the one-minute lag it would take before the diesel backup system became fully operational. So as not to disrupt power production, the test was planned at a time to coincide with a routine shutdown for refueling. However, fuel at the end of its cycle is most enriched in the fission products, making this reactor particularly unstable and vulnerable. Additional circumstances exacerbated the situation, and the system went out of control. Disaster occurred when a large amount of steam was generated in an explosion as cooling water came in contact with the hot core rods. A second explosion ensued when hydrogen produced by the reaction of steam with hot metal combined with oxygen. Because the facility did not have a containment structure, the explosions ripped the roof and released a large amount of radioactive material. Oxygen then came in contact with hot graphite, and the burning graphite contributed to the release of more radioactive material. The immediate surroundings of the plant were badly contaminated. The crops and farm animals had to be destroyed, and even after 20 years this area remains uninhabited.

The explosion itself was not a nuclear event *per se;* it resulted from the melting rods coming in contact with cold water. Prevailing winds dispersed the radioactive material in the next few days over a very large area in northern Europe covering parts of Ukraine, Belarus, Sweden, Russia, and many other countries. Plant operators who were at the site and the rescue workers who went in immediately following the accident were exposed to massive amounts of radiation (probably several hundred rem); 47 died from the exposure, and many more were stricken with radiation illness. An additional nine died in the ensuing months from acute radiation exposure, and 231 suffered from radiation illness.

Because many of the isotopes were short-lived, the radioactivity declined to half its value sharply within the first 15 minutes. After three months, the radiation level was less than 1% of the initial value. About 600,000 people received low-level radiation from the plume that spread over much of Europe. A 2005 report by the International Atomic Energy Agency (IAEA) places the number of additional cancers from this exposure at 4,000.[30] As large as this number is, it is much smaller than the 120,000 cancer cases estimated to occur in the same population during their lifetime. Against this large background, it is hard to identify which incidents of cancer, except perhaps the rare kinds like cancer of the thyroid, are direct results of exposure to radioactivity from the Chernobyl accident. In a similar vein, based on the general cancer rates, about 20,000 cases would have occurred among the 100,000 survivors of the nuclear weapons explosions in Hiroshima and Nagasaki. Based on the estimated radiation exposure (less than 25 rem), the number of cancer cases arising from fallout radiation would have been around 1000.

Explosion

The mental image that some people have of a nuclear reactor accident is like the mushroom cloud explosion of a nuclear bomb. If a major accident occurred, most likely from a loss-of-cooling event, it would not cause a nuclear explosion because all current reactors are configured to make a nuclear *explosion* extremely unlikely, if not impossible. As demonstrated by the Three Mile Island incident, almost all of the radioactivity from the reactor would be held within the protective containment vessel. As noted, the explosion at Chernobyl was primarily from a steam explosion, and this would not occur under present operational procedures.

Following the Three Mile Island accident, the NRC addressed probabilities for (1) escape of harmful radiation, (2) effects of that radiation on people living near the emission, and (3) damage to the surroundings from the spread of radiation to buildings and agricultural lands and waters. The NRC then applied probabilistic risk assessment techniques to the U.S. nuclear

30. Chernobyl Forum. *Chernobyl's Legacy: Health, Environmental, and Socio-economic Impacts and Recommendations to the Governments of Belarus, the Russian Federation and Ukraine,* International Atomic Energy Agency, Vienna 2003–2005.

stations considered likely to apply for operating license renewals that would permit their operation for an additional 20 years beyond the original 40-year permit.[31]

The probabalistic analysis in each case was not based on the worst-case scenario, which would have included complete discharge of the radioactive materials to the environment and a wind direction that carried all of the radioactive materials released toward the nearest and largest population center or vulnerable agricultural land. Instead, the analyses assumed that containment worked for the most part, and assumed probable outcomes for wind directions and wind velocities by sector. The study estimated the likely paths of any hazardous materials released, the quantities of that material as calculated from the quantities of the materials calculated to be present, and their toxicity to humans. Those data were applied to the population density and infrastructure in each of 16 sectors surrounding the plant to arrive at estimates of probable effects.

Exhibit 6.9 presents some representative results for 10 of the stations studied by the NRC. As indicated, threats to human health and life are greater for stations near population centers in the vicinity of buildings and industrial activities. For example, the Palo Verde reactor located about 60 miles west of Phoenix in rural Arizona with a capacity greater than $3\,GW_e$, has a predicted latent death count of less than 2 over 60 years. In contrast the two reactor stations at Indian Point station, 24 miles north of New York City, with a total capacity of $2\,GW_e$ have a combined projected fatalities from additional cancer cases of 50 over the same 60-year period. Note that the NRC study results imply that no immediate fatalities will ensue from radiation or radioactivity releases for those living or working outside any reactor with a containment vessel. The Union of Concerned Scientists in a recent report criticized the probabilistic risk assessment the NRC used because of its assumptions that the containment system would work and that multiple simultaneous cooling pipe breaks were unlikely.[32] Furthermore, the studies did not assume containment failure such as that possible in the event of a missile strike or the crash of a fully loaded large aircraft. (The NRC is currently studying these possibilities but considers them highly unlikely.)

If instead of the probabilistic approach we consider a worst-case scenario for the Indian Point plant and assume that the entire population of about 11 million was exposed instead of only 2.9 million in the downwind sector, the latent fatalities over 60 years from additional cancers would rise to 200—a number significantly higher than 50 estimated by the NRC but not astronomically so.

31. Nuclear Regulatory Commission. *Generic Environmental Impact Statement for License Renewal,* NUREG-1437, Washington, D.C., 1996.

32. L. Cronlund, D. Lockbrand, and E. Wyman. *Nuclear Power in a Warming World,* Union of Concerned Scientists, Cambridge, Mass., 2008.

Exhibit 6.9.

NRC Estimates of Fatalities and Property Damage Resulting from an Accident at Various U.S. Nuclear Power Operations

Station	Location	Capacity (MW)	Affected Population (Thousands)	Total Fatalities over 60 Years[a]	Percent of U.S. Population	Estimated Damage ($ Million)[c]
Braidwood	IL	2,397	1,615	20	0.0012	51
Brunswick	NJ	1,910	254	3	0.0012	11
Calvert Cliffs	MD	1,690	1,459	19	0.0013	43
Grand Gulf	MS	1,250	268	6.9	0.0026	14
Indian Point	NY	2,069	2,864	50	0.0017	43
McGuire	NC	2,320	890	9	0.0010	14
Millstone	CN	2,040	154	20	0.0013	46
Palo Verde	AZ	3,998	690	1.6	0.0002	5.2
Pilgrim	MA	690	486	3.2	0.0007	16
San Onofre	CA	2,150	1,284	15	0.0012	27

Estimates are based on NRC probabilistic risk assessments in *Generic Environmental Impact Statement for License Renewal*, NUREG 1437, Washington, D.C., 1996.

[a] Total estimated fatalities caused by the expected radiation exposure within the affected population for the 60 years following the accident.

[b] Damage to structures, loss of revenue, etc., in 1994 dollars.

Proliferation: The Global Scale

Proliferation among countries has long been a concern of the signatories of the Nuclear Non-proliferation Treaty), which include the United States, Russia, China, and most other nations using nuclear power. Specifically, their goal is to prevent other countries from acquiring the technology necessary to make a nuclear bomb. This effort has had only limited success, due to the acquisitions of nuclear weapons in recent decades by India, Pakistan, Israel, South Africa, and North Korea.[33] Current concerns center around North Korea and Iran, two nations with nationalistic ideologies that are at different stages of developing military nuclear technology.

Another concern is the possibility of nuclear waste being used by terrorists to create either an atomic bomb or a "dirty bomb" that causes radioactive damage. The damage that a dirty bomb can really create is one of contaminating a vital area so as to cause mass disruption. A dirty bomb will also cause a small increase of cancer risk for the population as a whole, but it does not lead to immediate loss of life. Terrorists are more likely to try something more spectacular, including conventional bombs.

33. South Africa has destroyed its few weapons and is no longer a member of the nuclear weapons club.

The hurdles for making a dirty bomb are also significant for a terrorist organization. For starters, amassing waste radioactive materials from power plants without killing themselves before they can make the device is an insurmountable challenge for the terrorists. While the radiation from the dirty bomb *after* it has been exploded—and the radioactive materials dispersed—poses only a small risk to the large population, in a concentrated form, as would be necessary for making the bomb, the acute radiation exposure to those assembling the device will be high enough to cause illness and death. The terrorists could decide to make a dirty bomb from alpha particle emitters, because they can protect themselves against this nonpenetrating radiation. For this variety of dirty bombs, the radioactive materials can be obtained from a variety of medical and other devices that have nothing to do with nuclear energy, so this threat is not reduced by turning off nuclear power plants.

The risk of a terrorist group acquiring enough fission material to make a small nuclear bomb from nuclear waste is a different question. We must distinguish between nuclear devices based on uranium and those based on plutonium because the two pose different hurdles. The uranium device is easier to make. It is based on a straightforward gun design that shoots one subcritical mass of uranium into another subcritical mass. The pieces together exceed the critical threshold, and a nuclear explosion ensues. However, getting the uranium fuel and enriching it to weapons grade is the difficult step, as we discussed earlier. As for the plutonium-based device, acquiring the fuel is relatively easier. Plutonium can be chemically extracted from spent nuclear fuel if one has access to it. However, the difficult step is making a plutonium bomb. In order to successfully explode a plutonium bomb, it is important to keep the ^{239}Pu from poisoning itself from its own neutrons. To achieve this, plutonium is fashioned into a thin-walled sphere so that the neutrons mostly escape and are not absorbed by the neighboring Pu atoms. In order to explode the device, the Pu sphere has to be compressed together into one lump. This feat is achieved by a process of implosion. A number of chemical explosives are set off around the Pu sphere, and their shock wave compresses the sphere. The positioning and timing of the explosions has to be very precisely controlled, or the nuclear device simply fizzles. The sophistication and precision involved in making a plutonium device make it difficult even for countries with significant resources to reliably succeed.

There will always be dissident groups seeking to call attention to their grievances by violent means, and therefore the risk of having such a group acquire and deploy a small fissionable bomb exists. While this risk could be minimized with adequate protection of important secrets and technology, the real risk could come from a rogue nation providing a terrorist group with one or more small nuclear devices and a means for delivering them. Even so, a one-kiloton "suitcase" device does not wreak as much damage as many conventional non-nuclear explosive charges do.

We do not mean to take the threat of terrorist attacks using a nuclear device lightly, for it can cause mass economic disruption. Ours is not a situation with easy choices, but the growth of nuclear power is important for the energy security of the world. This is primarily not a technical problem but one that needs a political solution. Greater involvement of the IAEA in monitoring of national nuclear programs and internationalization of the enrichment and fuel reprocessing are proposed by Mohamed ElBaradei, director general of the IAEA and winner of 2005 Nobel Peace Prize. Internationalization of fuel processing could provide assurance to countries pursuing nuclear power that they will get the fuel for their plants, and at the same time assure their neighbors that nuclear material is not being diverted to weapons development. Stable energy supplies can be a deterrent to war. By having personnel from many different countries supervising the fuel processing, internationalization provides the safeguard that nuclear material is not being diverted to military uses.

Disposal of Nuclear Wastes

Radioactive wastes take many forms and are categorized as low, intermediate, or high level according to the intensity of their radioactive emissions and the period over which that intensity persists. Many radioactive isotopes used in medical diagnosis and treatment are short-lived, and their wastes can be easily stored at their points of use for the days or weeks needed for them to decay to essentially undetectable levels of radiation. Other useful radioactive substances like ^{60}Co present greater control and disposal problems. The useful life of a ^{60}Co device in a medical or industrial application may be as long as 20 years, but this isotope will not be ready for disposal as low-level waste for at least 100 years. Most low-level radioactive wastes are placed in drums that are sealed and transferred to commercial disposal sites that operate under state control and follow federally mandated standards. The United States has relatively few of these sites, and in some states existing sites are required by state law to refuse to accept wastes generated in other states.

The problems involved in the disposal of low-level wastes pale in comparison to those associated with disposing of the high-level actinide wastes. (The 15 actinide elements encompass atomic numbers 89 through 103—actinium through lawrencium, and include uranium and plutonium.) All of the actinide elements produced in both civilian nuclear power operations and the production of nuclear weapon generate radioactivity that has the potential to do harm for many thousands of years.

During the years that production of nuclear weapons was considered a national priority, limited thought was given to what were viewed as ancillary programs, including disposal of weapons production wastes. In the United States, fluid effluents from the plutonium extraction processes were stored in steel tanks (which were sometimes placed in shallow concrete ditches), and radioactive solids were often buried in trenches. In the USSR some wastes were disposed of in open ponds. The major assurances cited for these methods were the remoteness of the sites and the absence of all but plant workers at those sites. However, with the passage of time, the steel tanks have been found to leak, weapons manufacturing sites have been closed, and, more important to the future of nuclear power, such "open disposal" methods have been discredited.

Over the same period, civilian nuclear power production has been creating quantities of high-level waste that now exceed the waste from weapons production. As a result, the problem of disposal is more pressing, more controversial, and seemingly more difficult to solve. When discussing the handling of radioactive waste, we should remember that the amount of radioactivity in a given sample does not change when the sample is spread over a larger area. While the amount of soil (number of tons) to be treated may increase, the total radiotoxicity—the deleterious effect on living matter, specifically mammals, that an ingested radioisotope can produce—does not increase. If anything, the radiotoxicity decreases as time passes and the radioisotopes decay to produce stable ones.

The amount of waste from nuclear plants is not particularly large. Globally, there are roughly 300,000 tons of high-level waste, which consists of spent fuel and waste from reprocessed spent fuel. This amount could fit in a two-story building covering the area of a typical city block. Each year about 13,000 tons of additional waste are produced. Most of the high-level waste is currently stored at the nuclear plants, either under water or in concrete structures. Because the volume of the waste is relatively small, it has not been a major concern thus far, which has given us time to find suitable sites for final disposal. Now, within a decade, there will be space limitations at the nuclear sites, and the waste will have to be relocated to a permanent or a semi-permanent storage facility.

Opponents of nuclear power have used statements to the effect that it takes hundreds of thousands of years for the radioactivity of spent fuel to reduce to background levels and questioned whether it would ever be possible to safely store the waste. How can we assure that it will not leak out during that long time period and contaminate groundwater? The question deserves a closer look, and as we shall see, we do not have the technology to *completely* remove this risk. However, if the objective is to reduce the risk to some *acceptable* level, then the problem can be solved. Therefore it is important that we come an agreement on what should be the *acceptable* degree of risk, something that has truly not been resolved.

Wastes from nuclear power production include both spent fuel and the radioactive components of any plants that have been decommissioned. The spent fuel has a high level of radiation and is the one we are most concerned with. Spent fuel contains the unconverted uranium and also fission products, such as cesium-137 and strontium-90, which are highly radioactive with relatively short half-lives of around 30 years, and also moderately active nuclei such as ^{239}Pu and other higher actinides with very long half-lives (24,000 years or more). The high radioactivity also means that the material can get very hot, and for that reason, the spent fuel is kept onsite and under water for many years to let the short-lived isotopes decay down to more stable products. After

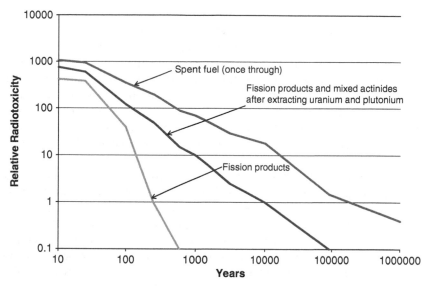

Exhibit 6.10. Decay of radiotoxicity of spent fuel over time

a few decades under water, the spent fuel is moved out of the water and stored in concrete blocks. In a hundred years, the radiotoxicity of the spent fuel is reduced to about 1,000 times that of the uranium ore. For its radiotoxicity to decay to the level of uranium ore, it would take more than 100,000 years. The key question is how should this material be stored, and what measures can be taken to mitigate the risk of it leaking out of the containment and into our drinking water supply.

A study conducted by the International Commission on Radiological Protection examined the decay in radiotoxicity of spent fuel as a function of time. The data from that analysis are shown in exhibit 6.10. The graph shows the relative radiotoxicity of the spent fuel and also breaks it down into the contributions of the short-lived fission products, long-lived actinides, and the toxicity expected from removal of actinides by reprocessing. Please note that the plot is in logarithmic and not linear scales, and thus is compressed at high values and expanded at lower values.

Looking at the line representing the total radiotoxicity in exhibit 6.10, we see that after 10 years the spent fuel is about a thousand times more toxic than the uranium ore (relative radiotoxicity of 1), and the toxicity will decline to the level of uranium ore in about 130,000 years. If uranium and plutonium are extracted from the waste, as is the practice with reprocessing, the radiotoxicity decreases more steeply. If all the actinides are extracted, the radiotoxicity of the remaining fission products shows an even steeper decline, and it reaches the level of the natural uranium ore in less than 300 years. Thus, if the longer lived plutonium and thorium (along with curium and americium) in the spent fuel were to be extracted for use in breeder reactors, managing the remaining spent fuel would be greatly simplified.

The benefits of reprocessing are first that the unconverted uranium and plutonium are recovered for use as fuel and thus minimize the need for mining more uranium. Second, and more important, the volume of the waste is reduced to less than a fifth and thereby made more manageable. Furthermore, because the remaining waste contains primarily highly radioactive fission products whose toxicity would decay to the background levels in about 300 years, it makes storing and safeguarding this waste technically much more feasible.

Measures such as immobilizing the radioactive elements in a glass, and storing that glass in steel containers that have been protected against corrosion (by the use of sacrificial copper or electric current in a process known as cathodic passivation) can reliably prevent leakage for a hundred years or more, and it does not seem a stretch for us to be able to safeguard a repository of such materials for 100–200 years. We should also remember that the radiotoxicity shown in exhibit 6.10 refers to a hypothetical case when all the material is actually ingested by the population. The probability of leaking the entire contents is extremely low. One can imagine a few percent of the material leaking out, and only a fraction of that amount would actually get into groundwater and be consumed. Thus, radiotoxicity from ingestion of 1% of the waste would equal that from natural uranium ores when its relative toxicity is a 100 times higher. The line for the waste reprocessed fuel reduces to 100 times the baseline level in less than 200 years. We have to compare this risk against the fact that natural rivers, like the Colorado in the western United States, are draining larger amounts of natural uranium-bearing ore and its decay products, and the calculated risk from them is many times larger.

Two questions that we have to ask are first whether reprocessing the fuel to extract plutonium and thorium presents an unacceptable risk of nuclear proliferation via subterfuge, and second, whether the additional risk of storing spent fuel is greater than the downside of not including nuclear energy in the portfolio. In light of the new storage technologies, discussed below, it seems quite reasonable that we will be able to safely store the spent fuel for a few hundred years. On the other hand, if we choose not to develop nuclear power further, it would entail that developing economies would either have to use even more coal (i.e., more pollution, greenhouse gases, and direct health hazards) or deny their population the emergence into the world market because the alternative income sources of energy are either not ready or too expensive.

Options for Waste Storage/Disposal

The largest amounts of high-level nuclear wastes are currently found in countries that have had large nuclear weapon or nuclear power programs. As a large generator of both peaceful and weapons-related radioactive wastes, the

United States has paid considerable attention to waste disposal. Although it chose in 1976 to forego reactor fuel reprocessing to recover useful plutonium and uranium, the United States has not been successful in convincing other nations to stop reprocessing. Unlike the United States, Britain, France, Japan, and Russia have been treating the spent fuel from their nuclear power plants, but they have apparently not yet decided on methods or locations for disposing of high-level civilian or weapons program wastes. Both Britain and France have operated facilities to immobilize the wastes from spent fuel reprocessing, and France has reprocessed waste from Japan.

Several potential ways exist for storing wastes from nuclear power production and from previous nuclear weapons production, or for rendering them relatively innocuous. Among those that have been suggested are the following:

Waste disposal in secure underground facilities

Waste disposal in secure facilities below the seabed in deep ocean locations

Dispersal into space using rockets

Treatment in special accelerators to destroy the most radiotoxic and long-lived isotopes in the waste, also known as transmutation

Of these methods, underground disposal is preferred. The use of the seabed would present many territorial, diplomatic, and political problems; dispersal by rocket would carry too high a probability of failure, even if only one in a million; and the effectiveness of the transmutation scheme is not proven and in any case would likely be economically prohibitive.

The United States, Finland, Sweden, and Switzerland are exploring underground disposal options. The United States now operates a facility that permanently stores actinide wastes that were generated during weapons and power production as well as during research and development projects. Sweden, Switzerland, and Finland are considering the disposal of their discarded nuclear fuel elements by covering them with copper and encasing the resulting matrix with stainless steel. The sealed package would then be placed in a hole drilled to 1,000 ft or more below the surface in a competent granite formation free of water, and then covered with water-impenetrable clay.

Other countries with substantial quantities of waste from nuclear generated power and weapons programs (e.g., China, France, India, Japan, the Russian Federation, South Korea, and the United Kingdom) are apparently less advanced in choosing disposal options.

The Current U.S. Disposal Agenda

Although regulators were aware of the need to safely dispose of the high level wastes as early as the mid-1950s, it was not until 1982 that the U.S. Congress passed the Nuclear Waste Policy Act. And it was not until 1987, in a modification of the Act, that Congress selected Yucca Mountain in Nevada as the repository for high-level nuclear power and weapons-related wastes. As modified, the Nuclear Waste Policy Act requires the U.S. Department of Energy to build and operate the facility, the NRC to approve the construction plans and proposed site operations, and the U.S. Environmental Protection Agency to certify that the proposed operations would not affect populations in the surrounding area. As of this writing the fate of the chosen site in Nevada is again on hold while the Obama administration reviews its suitability.

The majority of wastes to be stored at Yucca Mountain will be spent commercial nuclear reactor fuel, with waste from nuclear weapons processing and naval reactor fuel constituting the remaining portion. The spent fuel will arrive, either directly from reactor sites or from interim storage facilities, in shielded casks consisting of individual fuel rods or perhaps clusters of rods. The casks or clusters will then be transferred to the facility's storage containers (described below). When a storage container is fully loaded, it will be conveyed to storage tunnels drilled in the mountain.

The proposed plan for the Yucca Mountain storage facility relies on the following engineered safeguards:

The storage containers will be made of a thick corrosion-resistant stainless steel. The containers will be mounted on V-shaped pallets made of corrosion-resistant steel positioned in the facility's tunnels. Mounting on the pallets will help reduce the probability of the containers' contact with standing water and ensuing corrosion. (Even in the continual presence of moisture, the primary containers are estimated to remain intact for 10,000 years or perhaps much longer, but this estimate cannot be proven in our lifetimes.)

The containers will be further protected from corrosion by a titanium shield placed over them that protects them against water dripping from above.

The tunnels will be air conditioned to remove heat from the decay of the radioactive elements in the containers; reduction in heat at the container's surface will help lower corrosion rates, and flowing air will help remove moisture.

The plan calls for all of the tunnels to be kept open for 300 years before closure to allow for detection and rectification of errors in filling and moving the containers to their final location.

Despite the engineered safeguards listed above and the careful facility design, the selection of Yucca Mountain may be an excellent example of the truth of the adage "look before you leap." Yucca Mountain's geology is not ideal, as it is characterized both by "welded" and "unwelded" tuffs (a tuff is a rock compacted from volcanic ash, varying in size from sand to course gravel). Welded tuff is characterized by fractures through which water and associated dissolved radioactive materials could move with relative ease; unwelded tuff is more porous but more resistant to water flow. The storage site also rests above the water table that serves the surrounding area. This factor and the site's lack of geological homogeneity make its long-term integrity for storage difficult to quantify sufficiently to address critics' concerns. Extensive studies have been conducted to judge the potential for material to (1) escape storage containers, (2) then escape the facility in the event that the containers become corroded, and (3) once escaped, to reach the surrounding groundwater and eventually the people and farming activities that use this water.

Many additional safety-related issues have arisen, triggering debate. Could the site be damaged by a volcanic eruption? Some eruptions occurred in the area several million years ago. Could climate change result in much higher rainfall levels at the site instead of the current 6–7 inches/yr? Could earthquakes cause the tunnels to collapse? There is evidence that earthquakes occurred long in the past. Might they occur again and thus change the site's geography in significant ways? If water reached the containers storing the waste, how long would it take for them to corrode? If the containers corroded and all the safeguards failed, how long would it take for water and the radioactive contents dissolved and leached from the containers to reach the underground streams and carried to drinking water and irrigation wells? If leaching occurred, how soon would it take place, and what would be the amount of the released radioactivity? As shown in exhibit 6.10, if unprocessed spent fuel was stored, little harm could come from its leakage after about 1,000 years but potential hazards may continue to exist for 100,000 years or longer. With fuel reprocessing the radiotoxicity of waste, which arises primarily from the fast-decaying fission products, reaches background level in about 300 years.

In addition to the Yucca Mountain facility, which is far from being operational, the Waste Isolation Pilot Plant (WIPP) in southern New Mexico near Carlsbad Caverns has been approved and has begun operations to store the actinide wastes. Like the Yucca Mountain facility, WIPP was authorized after considerable debate and scientific and technical analysis, including probabilistic risk assessment.[34]

WIPP consists of caverns mined in a small part of a salt bed that measures roughly 690 miles by 260 miles and lies under a layer of caliche (a water-impermeable rock/soil mixture) and over another impermeable rock layer. WIPP accepts actinide wastes of all kinds from 23 locations in 15 states, and shipments have been made of both low- and high-level wastes. Material to be stored is treated and processed at the surface as necessary, transported to the storage level, which is 2,150 ft underground, and placed in tunnels mined from the salt

---

34. This methodology was first applied to the analysis of reactor safety in the landmark study known as WASH 1400: Nuclear Regulatory Commission, *An Assessment of Accident Risks in U.S. Commercial Nuclear Power Plants,* Washington, D.C., 1975.

Exhibit 6.11.
Requirements for Producing 1 CMO/yr Nuclear Power

Activity	Facility Requirements (Units/CMO)	Environmental Impacts	
		Area (mi²/CMO)	Other
Mining			
Surface mine	500	1[1]	
Underground mine	1,000	5	Exposed uranium and daughter products
Ore concentration	200	3	As above, but better covered and more carefully monitored
Conversion/ enrichment	25	5–10	Residual depleted uranium as a fluoride or oxide
Fuel fabrication	40	<1	Negligible
Reactor operations (electricity production)	2,280	2,000–5,000[2]	
Reprocessing	40	2–4	Negligible
Waste storage			
Spent fuel and high-level waste	0.5	1.5	
Low- /intermediate-level waste	5 to 20?	10–30?	Leaks, spills

[1] Operation in waste lands; no restoration.
[2] Exclusion area, 0.5-mile radius; limited-access area, 1-mile radius.

bed. Over time, Earth's pressure will deform the salt, the tunnels will shrink, and the stored material will be permanently sealed in a water-free environment.

Requirements and Potential Impacts of Producing 1 CMO/yr of Nuclear Power

As with the use of fossil fuels, the number of facilities and areas associated with them for nuclear operations depend on the choice of specific technologies. To estimate the requirements for producing 1 CMO/yr of nuclear power, which are listed in Exhibit 6.11, we have assumed the widely used PWR technology in a once-through mode. If we further assume that power will be generated in 900 MW plants operating with a capacity factor of 85%, the number of power plants needed would be 2283. These plants would annually consume 50,000 tons of uranium fuel, which would require mining, milling and enriching about 350,000 tons of natural uranium, two-third of which are produced by underground mining and the remainder by surface mining. If the uranium were present at 0.2% concentration in the ore, it would mean mining 175 million tons of ore, producing more than 500 million tons of tailings at a typical 3:1 ratio for earth moved to ore recovered. If the tailings were piled 50-ft high, the area covered would be roughly 10 mi^2.

The area required for the nearly 2300 power plants themselves would be about 125 mi^2, but each plant is required to have an exclusionary or limited-access zone to protect people and animals from accidental exposure in the case of a failure in containment. This zone extends for a mile or more around the plant depending on the terrain, meteorological conditions, and local regulations. In the United States, the limited-access zones are often for several miles. The Nuclear Regulatory Agency has published data for 113 plants at 72 sites in the United States.[31] Together these plants produce 0.054 CMO/yr of electric power, and they encompass 5,000 mi^2. If this proportion held for the rest of the world, producing 1 CMO/yr of nuclear power station would require 90,000 square miles of land. This is an extreme case and we note that many countries, including those in Europe, have less stringent requirements.

Conclusions

Nuclear power is one source of primary energy that has the potential to deliver large quantities of essentially CO_2-free energy at relatively low cost. Nuclear plants operate round the clock with very little downtime. They are suitable for base power and therefore would be most able to meet future increases in electricity demands. New technologies have been developed to increase the efficiency of power production and also improve the safety of nuclear power plants. Long delays in permitting and construction have been a major factor in increasing the cost of nuclear power, but having standardized designs could ameliorate these issues.

Global uranium reserves (those based at $59/ton) are only 11 CMO and could not last a long time if nuclear power use increased substantially over the current 0.2 CMO/yr level. However, the resource base could be increased more than ten-fold at higher prices for uranium ore, and this increase would only marginally increase the cost of power because fuels costs are only 15% of the cost of nuclear power. The supplies could be further augmented if we implement breeder technologies and/or reprocess spent fuel. Reprocessing reduces the amount of nuclear waste, and the waste remaining after reprocessing decays to background levels in about 300 years, thus rendering it much more amenable to storage than unprocessed waste.

The future of nuclear power as a contributor to world energy needs is likely to depend as much on public attitudes—and resulting government actions—as on its competitive economic position. Although several recent studies indicate that large nuclear power installations produce much cheaper electricity than do coal- or gas-fired power units, in many nations nuclear power continues to face opposition based on public apprehension about its potential harmful impacts as a source of ionizing radiation and thus a source of cancer and other deleterious health issues.

Nuclear power is hardly a perfect technology, but it is constantly improving in efficiency and safety. With some adjustments, it could become an even more crucial source of energy, as demand grows worldwide. Our discussion of the Chernobyl disaster, the likely outcome of other nuclear accidents, and the dangers from nuclear waste leaks during storage should make it clear that the *perceived* fears of nuclear accidents as portrayed in the media and by some opponents are grossly exaggerated. Nuclear power is not without risks, but neither are other sources of energy. We have to compare the real risks of nuclear power against those posed by other sources, including coal and the income energy sources (or renewables). In chapter 7 we discuss the income energy sources. The income sources have fewer associated hazards, but they currently provide only a very small fraction of our primary energy. As we shall see, when we try to scale them to a CMO level, significant environmental issues arise even with these "green" sources.

7

Our Energy Income

Geothermal, Hydropower, Wind,
Solar, and Biomass

We briefly discuss our income resources, or renewables, as they are often referred to, in chapters 1 and 3. These resources differ from our inherited resources in that they can potentially last for as long as civilization exists and beyond. We would be better off if we could live off our income, instead of relying on our inheritance that will some day be exhausted. While historically we survived on income resources for many millennia, those sources cannot support our current lifestyle, and they contribute only a very small portion to our total annual energy budget. Given the overall desirability of switching to income sources, in this chapter we review the status of different technologies for using them and what it would take for each one to become a substantial contributor. As we shall see, none of these technologies is currently economic, and they are not free from being potentially damaging to the environment when grown to the required scale. Income resources can offer a path toward sustainability *if* we are able to engineer the systems correctly without creating other problems along the way. With that in mind, we begin by enumerating our principal energy income resources:

- Radiant energy from the sun, which also drives the wind and the water cycle, and provides the energy for plants to grow
- Heat energy from Earth
- Tidal energy derived from the moon's gravitational attraction
- Wave and ocean-current energy

The radiant energy derived from the sun is by far the largest contributor (see chapter 3, box titled "Total Solar Flux"). It comes from a series of solar reactions that result in the fusion of hydrogen into helium. The sun's radiant energy directly or indirectly spawns biomass, photovoltaic electricity, solar-thermal, and wind energy. It also results in derivative energy sources such as wave energy (from wind), ocean-current energy, and hydropower. In chapter 3 we briefly introduced these income resources and noted that the amount of solar radiation reaching the surface of Earth is around 23,000 CMO/yr. This very large and potentially usable amount is divided into direct radiant energy (about three-fifths), the water/evaporation cycle (about one-third), the air-motion (wind) cycles (about one-thirteenth), and a photosynthesis component, which accounts for only a minute fraction of the total solar contribution (about one-thousandth) but that is essential to life.

Yet, the income resources we currently use add up to less than 0.35 CMO of energy. This utilization is mainly from hydropower and biomass (wood burning). Wind and direct solar account for less than 0.005 CMO, and all biofuels combined contribute around 0.01 CMO. Today solar energy is used primarily indirectly, in the growth of crops that produce fiber and food, and in the natural growth of timber—although some of the latter is produced through plantation farming. Some crops can also be used as sources of heat and fuels, as can timber. While these contributions are now small in comparison to world commercial energy production, they are expected to grow in the future. Even a sizable future contribution of biomass to our energy needs could well be exceeded by future use of direct solar heat and solar derivative sources, such as wind power.

In this chapter we consider income resources in terms of their distribution and overall potential for providing a significant portion of global energy needs in the future. We estimate the current status of their use, and describe relevant technologies under development and what might be needed to increase their utilization to a CMO scale. We concentrate on commercial, larger scale uses, as these would have the greatest influence on global energy production for many decades. We do recognize that small-scale uses can greatly change the lives of individuals, particularly those living in the villages of developing countries. Often these villages are not served by the existing electrical grid, and it would require substantial investments in power generation and transmission capacity to get them connected. Distributed power generation systems using income resources would avoid the expense of transmission. If it were affordable and of a size that matched the relatively low demand of those villages, it go a long way in improving the lives of the people and buying time to develop large-scale economic systems based on income sources.

As we review the different income sources and the various ways to use them, keep in mind five questions. The first is the amount of extra energy that each source or method provides, and whether it adds to the total available energy pool at a scale capable of making a global difference. Second, and closely linked to the first, is the availability of the resource—if it is unavailable over certain periods, will we need to back it up with fossil fuels? The third is whether the resource avoids the emission of greenhouse gases in its production and use. The fourth question has to do with the energy security the source provides for the nation that uses it. The last question, fifth, is, of course, the price. The best energy resource in the world on other fronts would not mean anything if we could not harvest it cost-effectively.

Geothermal Energy

Resource Base

Geothermal heat comes from Earth's general cooling and gradual freezing of its now molten inner core. The core is also rich in heavy radioactive metals including isotopes of uranium and thorium, and their daughter isotopes, that Earth acquired when it was formed some 4.5 billion years ago. The radioactive decay of those elements provides additional heat to the core, but even without this contribution, the core cooling could continue to release an estimated 7 CMO/yr for another 4 billion years.

Geothermal energy today is obtainable mainly in those regions of the world where the continental plates are colliding. This collision results in lifting of the hot, semisolid material at the interface of the core and mantle to crustal locations near Earth's surface. A "rim of fire" that covers the areas on the western coast of the Americas and some parts of the Asia Pacific region, as well as a few more spots—for example, Iceland and Italy—are primary regions where geothermal energy is readily available. While volcanoes are the visible manifestations of this magma uplift, extraction of geothermal energy from a site near an active volcano is not practical. Instead, most of the extraction of geothermal energy in the world today is mediated by underground water reservoirs that are heated by their proximity to the magma. This resource is currently used for power production or providing heat. In places where the water temperature is higher than about 300°F, production of electric power is feasible, and where the water temperature is hot but less than 300°F, the water can be used for direct heating. Direct heating requires an infrastructure to distribute the hot water to the place of use, which cannot be too far from the source.

Other geothermal resources, representing a considerably larger resource base, include geopressurized brines and geothermally heated dry rocks (HDRs). Development of these resources would require injection of fluids to extract useful heat. HDRs are generally located 5–10 miles below Earth's surface, but technology to access them economically needs further development before it can be ready for commercialization.

Power Production

The water in geothermal reservoirs is under pressure and can be at temperatures as high as 360°F (~180°C), which is substantially higher than the boiling point of water at atmospheric pressure. When this water is brought to Earth's surface, it boils. The steam is either piped to provide heat for residential and commercial buildings or is used to drive turbines for generating electric power. Because these high-temperature waters are also often highly saline and corrosive, they are not used directly. Instead, a secondary fluid—often water—is used to extract the heat. In some cases, the hot water is used to first vaporize another liquid with a lower boiling point (often isobutane), and those vapors are used to drive a turbine. Finally, there are also HDR resources where there is no natural water table. In such cases, surface water must be injected deep into the HDR to extract heat. HDR resources are more widely located and represent a larger resource base, but so far it has not been economical to extract power from this resource.

Water originally present near magma can be used to transfer heat from magma to the point of use. With time, the water can become depleted, and the energy production at a specific site can decrease with time. For example, this has occurred at Larderello, Italy, where energy extraction in the 1970s had decreased to about 50% of the level in the early 1900s. The Geysers facility in Northern California suffered a similar but quicker decline in electric power output, which has been restored by injecting water from nearby wastewater treatment facilities in Santa Rosa.

Examples of major installations of electric power from geothermal sources (i.e., those greater than 30 MW) can be found on all continents except South America and Antarctica. There are plants in Kenya, the Philippines, Indonesia, Australia, New Zealand, Russia, the United Kingdom, Iceland, Mexico, Costa Rica, and the United States. The largest operation of geothermal power is in Geyser, California, about 70 miles north of San Francisco, where Calpine operates 19 plants with a combined net output of 750 MW$_e$.

Today, geothermal energy used to produce electricity probably accounts for no more than 0.003 CMO/yr. Its use for direct heat may be about the same as the electricity-production figure, for a total of about 0.006 CMO/yr. Bundschuh and Chandrasekharam provided a review of the various estimates for geothermal energy.[1] Karl Gawell estimated the potential of global geothermal power production from *known* sources to be around 1,000 TWh/yr, which translates to 0.06 CMO/yr.[2] An estimation by Bjornsson et al.[3] based on the distribution of volcanic activity places the global potential of geothermal energy at about 0.78 CMO, or roughly 12 times higher than the preliminary estimate by Gawell et al. The estimate of the low temperature resources, which are not

1. J. Bundschuh and D. Chandrasekharam. "The geothermal potential of the developing world," in J. Bundschuh and D. Chandrasekharam, eds., *Geothermal Energy Resources for Developing Countries,* Swets and Zeitlinger, Lisse, 2002, pp. 53–61.

2. Gawell, K., Reed, M. and Wright, P.M., *Preliminary Report, Geothermal energy: the potential for clean power from the earth,* Geothermal Energy Association, Washington, DC, 1999.

3. Bjornsson, J. The potential role of geothermal energy and hydro power in the world energy scenario in year 2020. *Proceedings of the 17th WEC Congress,* 1998.

suitable for power production, is even higher at around 3.8 CMO. These lower temperature geothermal sources are located in common soils or rocks only a few feet, or a few tens of feet, underground and can be used to extract energy for heating buildings.

As we shall see, there are few income resources that can scale to a CMO level. Getting half a CMO from geothermal power would make a significant contribution in our switch to income resources. Power generation from geothermal resources can be increased substantially over the current rates, perhaps 50- to 100-fold. In most locations the estimated cost of geothermal power is between 8¢/kWh and 15¢/kWh, which is more expensive than wind power but still competitive with the retail price of electricity, particularly if "green" power is *priced* at peak power rates, a practice that some communities are adopting to provide financial support for developing our income resources.

The cost of upkeep for geothermal power is also low. Therefore, we can foresee a significant increase in the utilization of this resource. Geothermal energy is environmentally benign, and the primary concerns that expansion of geothermal resources is likely to encounter will stem from alternate uses of the land such as maintaining habitat for wildlife or infringing on the rights of the current owners. Additionally, geothermal developments could face resistance from the potential of increased seismic activity as is often encountered with drilling operations. Major expansion of geothermal energy will come only if new technology to extract useful heat from HDR is developed. In the interim, we must find our energy somewhere else.

Hydropower

Hydropower includes hydroelectric, wave, and tidal power, although the latter two technologies are not currently developed for commercial use. Hydroelectric power is a well-developed technology, used in several ways and sizes. It dates back to 1870, when the Cragside Dam in Rothbury, England, was commissioned. Since then more than 45,000 dams have been built worldwide. Many of them were built for the purpose of regulating water flows and avoiding floods, but many also were built to provide electric power. Helpfully, hydropower sites are more widely distributed than geothermal sites. Many rivers, or even creeks, can capture the energy of flowing water to supply mechanical and/or electrical power. While most flowing water can be used to drive water wheels, major production of water-driven energy is achieved only by impounding water behind dams of varying heights. By allowing the water to fall while passing through a waterwheel or turbine, the potential energy of the stored water is converted into kinetic energy that drives a turbine connected to a generator, which in turn produces electricity. The term "hydroelectricity" refers to the generation of electricity by hydropower. Before the invention of the electric generator, flowing water was used only for the mechanical power it could provide. The invention of electrical generators made possible the use of hydropower far beyond the site of the dam, falls, or rivers on which it depended.

Hydroelectricity is already a significant contributor to the world's energy needs. It furnishes 16% of the world's electricity, or 7.6% of our total primary energy on a fossil-fuel-equivalent basis. In most instances the power generated is fed into the local grid, although there are cases where the power is produced solely for use by certain industries. For example, electrolytic production of aluminum uses large amounts of electricity and tends to have dedicated hydroelectric projects associated with production plants because of the low cost of this power.

The cost of producing hydroelectric power is generally low because there is no cost for fuel and the installations tend to have long service lives, such that they continue to produce power even long after the initial capital investment has been paid off. Depending on the individual circumstance, the production cost of hydroelectricity ranges between 3¢/kWh and 12¢/kWh. The low operating costs, coupled with the fact that water can be stored behind the dam, means that utilities can use the facilities for either base or peak power production. In some cases, if there were surplus power in the grid, these facilities could also serve as pumped-water storage systems (see chapter 3). The capacity factors of most hydroelectric plants are around 50%, and the worldwide average is 46%. The annual variation in the capacity factor of hydropower plants is mainly a reflection of the variability in rainfall.

The primary requirements for hydropower are regions with abundant rain or snow and a geography that channels water (including snow melt) into streams that can be impounded. Locations best suited are where the water flows at the bottom of a canyon or gorge, and a high dam can be built. The mountainous regions of North and South America are examples of good sources of hydropower. Until recently, the United States, Canada, and Brazil were the nations producing large amounts of hydroelectric power, each with an installed capacity of between 70 and 80 GW. According to data from BP (formerly British Petroleum), in 2000 Brazil produced about 87%, Canada 60%, and the United States 7% of its electricity from hydro. In 2006, the corresponding percentages were 83%, 91%, and 7%, respectively, which reflects the recent growth of hydropower in Canada. There was a substantial expansion in hydropower capacity of the developing nations in the middle and late 20th century. Although some of this power is exported to nearby nations, hydropower is primarily a local/regional energy source, because electricity can only be transported effectively for about 600 miles over land, and considerably less if the transmission is through undersea cables.

The regional distribution of current and technically feasible hydropower production for 2006 is presented in exhibit 7.1. Commercial hydroelectric power sources currently contribute a total of about 0.17 CMO/yr of equivalent

Exhibit 7.1.
Regional Distribution of Hydropower Energy Production

Region	Current (2006) Amount (CMO)	Percent	Technically Feasible Amount (CMO)	Percent
Middle East/North Africa	0.014	7.8	0.018	2.5
Central/South America	0.034	19	0.18	25
Russian Group	0.014	8	0.066	9.2
North America	0.045	25	0.10	14
AsiaPacific	0.034	19	0.17	24
Sub-Saharan Africa	0.004	2.0	0.12	17
Europe	0.037	20	0.066	9.2
World Total	0.17	100	0.72	100

commercial energy. According to a report by the World Energy Council,[4] the total technically feasible amount of energy available from all water flow world-wide is 16 trillion kWh/yr, which converts to a little over 1 CMO/yr.

The Middle East/North Africa region produces negligible amounts of hydropower, with the sole exception of Egypt, which produces around 15% of its electricity from hydropower, mainly from the Aswan project. Currently, China has the largest installed capacity—almost 130 GW. Some regions (e.g., North America) have developed much of their available hydropower potential. In contrast, the Asia/Pacific region has substantially greater underdeveloped potential. There are many major hydroelectric projects afoot in China, India, Brazil, and Russia. In China alone, there are 12 projects with capacity of more than 3 GW slated for completion by 2015; together these 12 plants will have capacity of 90 GW. A full quarter of this generating capacity, 22.5 GW, will be from the Three Gorges Dam across the Yangtze River. The Three Gorges Dam is the world's largest hydropower project. It has already taken 15 years to build at a cost of $30 billion. With all of these additions, the world's total hydropower contribution could reach up to 0.5 CMO/yr.[5]

This projected near-term capacity might not be realized in practice, for several reasons. Current and future dams lose storage capacity from silting and debris accumulation. Also, as people have become more aware of the environmental impacts of dams, opposition to such projects has grown. Foremost among the concerns is the disruption and displacement caused to people living in the area, whose homes and land become submerged by the lake created

4. *2007 Survey of Energy Sources*, World Energy Council, London, 2007.

5. Estimate based on the assumption that actual production is 50% of installed power potential.

behind the dam. Thousands of people are displaced in the process. Not only do they lose their homes, but they often lose their livelihood that is closely tied to the land and find themselves displaced to an area that is foreign to the way they grew up. Their lives are altered along with the course of the river, and they justifiably feel wronged.

Compounding the humanitarian reasons, there are arguments opposing large hydropower projects based on closely linked technical and environmental feasibilities. In its report, the World Commission on Dams finds that while dams have made an important contribution to human development through flood control, irrigation, and power generation, they have in too many cases exacted an unacceptable price for those benefits by fragmenting and transforming the world's rivers and displacing an estimated 40–80 million people.[6] The lack of equity in distribution of benefits has also called into question the value of dams. The consequences of damming the river also include reduced flows downstream and attendant silting as well as loss of habitat for wildlife, including fish. These consequences affect an even larger area and more people than those only in the vicinity of the dam.

More recently, questions have been raised about greenhouse gas emissions associated with hydropower. Large quantities of steel and cement are needed for these projects, and their manufacture entails emissions of greenhouse gases. Estimates for these emissions are relatively straightforward to make and show that the consequences are small compared to the savings from greenhouse-gas-free power production afforded by the dam. The larger and more difficult-to-estimate greenhouse gas contributions occur during their operation. The submerging of large areas results in the microbial decomposition of vegetation, which produces substantial quantities of methane—a potent greenhouse gas. Furthermore, eliminating vegetation also eliminates the consumption of CO_2 by that vegetation. Estimating the amount of these gases is difficult, and amounts vary depending upon the specific location, terrain, and power intensity of the dam with its associated lake.

In chapter 6 we mention the Environmental Product Declaration documentation that power producers in Europe must file. According to the declaration filing by Vattenfalls, a Nordic company that operates nuclear, hydro, coal, wind, and gas-fired power generation plants, CO_2 emissions associated with the construction and operation of their eight Nordic hydroelectric power stations with a combined capacity of 8.6 GW amount to about 4.22 g/kWh. Only a tenth of the emissions are associated with the construction, and more than 75% of the emissions are a result of inundation of land. For comparison, we note that a typical coal-fired power plant, the most common alternative to power generation, produces approximately 1,000 g CO_2/kWh (2 lbs CO_2/kWh is a reasonable estimate for most coal-fired plants). We can expect

6. World Commission on Dams. *Dams and Development, a New Framework for Decision Making,* November 2000.

that inundation in the tropical areas would lead to submerging of much larger amounts of vegetation, and consequently lead to greater emissions of greenhouse gases. In keeping with that expectation, the World Council on Dams concluded that while "some values for gross emissions are extremely low, and may be ten times less than the thermal option...in some cases the gross emissions can be considerable, and possibly greater than the thermal alternative."

The Future of Hydroelectric Power

Hydroelectric power is a useful and relatively clean source of power, but there are clearly complications involved with some of its aspects, especially damming of rivers. There is a great deal of potential for growth within the industry, particularly with respect to micro hydropower (less than 100 kW). Micro-hydropower extracts power from the run of the river and does not require dams. These systems can be useful for supplying power to remote or underdeveloped areas that may be endowed with a flowing body of water. The main limitation of micro-hydropower is that the total amount of energy produced is not very large and is not likely to scale to a CMO/yr. Overall, though, hydropower will be an important and growing component of the world's energy system for a long time.

Wave Power

While dams are the primary source of hydropower, the waves of the world's oceans can also being harnessed for energy. The attraction of wave power is its potential to supply electrical energy to about half of the world's population living within 60 miles of a coast. Unlike wind, wave power is steady and can deliver power all of the time. There are many different concepts and machines for producing power from waves. None of these is cost-efficient for large-scale use at this point. The chief issues with wave power are that the systems necessary to extract power of even 1 kW require tons of machinery and cover acres of sea. The cost of electricity from wave power, estimated by considering the costs for installation and maintenance of the systems, is currently around 60¢/kWh, and if this power source is to be developed for delivering energy to coastal cities, then the costs have to be reduced drastically.

Tidal Power

Tides function on a consistent and predictable schedule, and thus like wave power, tidal power could be integrated with energy demands in a planned manner. Tidal power has the advantage of being able to extract more power than wave energy. Tidal power generators are built in areas with fast currents, preferably where the tide is concentrated into a small area by natural obstructions. The tide thus builds a high headwater pressure—enough to drive a water turbine. Most commonly, a barrage with sluice gates is built to fill the inlet during high tide and let the water out during low tide. Several countries are building or planning tidal power projects, including the United States, Russia, the United Kingdom, India, and Canada. Some contemplated projects are in the gigawatt range, and many others are relatively small. Large systems can cover miles of coastlines. Environmental effects include altering the pattern of coastal water flows and disrupting the sea habitat. Like large hydropower systems, tidal systems also tend to be very expensive to build. Because they produce power about 25% of the time, it takes a longer time to pay off the initial investments. For these reasons, tidal power does not at this time appear scalable to the levels needed to impact our global energy crisis.

Wind Power

Global Potential

Wind is created because the sun heats different parts of Earth unequally. The poles are much colder than the equator, and the continents heat and cool faster than the oceans. These temperature differences, coupled with Earth's rotation, give rise to different wind patterns and indeed all the weather that we experience. It is the energy in the flowing air that we try to harness with a windmill or a turbine. Winds are strongest in regions that permit unimpeded airflow. Thus, open oceans have the greatest theoretical potential for wind-energy generation, but they are not practical because we are currently unable to set up wind turbines out in the open ocean and have the power transmitted to the cities.

Generally, wind speeds of at least 10 mph are needed to make power generation practical. Western coastal regions in the Northern Hemisphere and eastern coastal regions in the Southern Hemisphere are generally favored as wind power producers. A global view of wind regions is presented in exhibit 7.2. This map, taken from Archer and Jacobson,[7] shows the annual mean wind speed at different stations measured at about 30 ft (10 m) above ground. Most wind measurements are made at that height. However, the new wind turbines with longer blades are placed more than 250 ft (80 m) above ground, where the wind speed should be appreciably higher. Recognizing this fact, Archer and Jacobson used data for 250 ft where such data was available, and they built

7. Cristina L. Archer and Mark Z. Jacobson. "Evaluation of global wind power," *Journal of Geophysical Research,* vol. 110, D12110 (pp. 1–20), 2005.

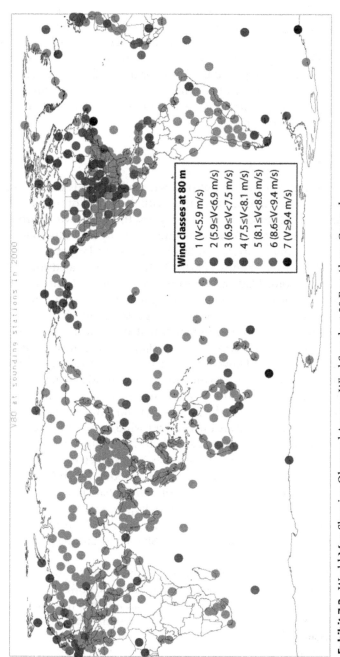

Exhibit 7.2. World Map Showing Observed Average Wind Speeds at 30 Feet Above Ground

a model to extrapolate the observed wind speed data to 250 ft level for all stations. They considered the area of the sites with winds of class 3 (see below) or higher, and also allowed for spacing turbines sufficiently far apart so as to not adversely affect power output. With this revision, they estimated annual global wind energy potential to be 72 TWyr of electric power, which is equivalent to 41 CMO. This quantity is an estimate of what is possible; it does not take into account competing land use issues or the challenges associated with accessing all those regions to install and extract power from the turbines. The analysis does show that, *in principle,* wind could provide more than enough energy to meet the world's needs.

Power Output of a Turbine

To a large extent, the power output of a windmill depends on the area that the rotors sweep. That circular area varies with the square of the rotor length, which is the radius of the circle. Over the years, the size of wind turbines has increased remarkably. In 1985, wind turbines typically had rotor diameters of about 50 ft (15 m) and a capacity of 50 kW. Today, the largest wind turbines have a rotor diameter of almost 400 ft (120 m) and are rated for 6,000 kW (6 MW). A longer rotor blade also means that the hub is located higher up, where the wind is not as affected by the roughness of the terrain and therefore tends to blow faster. A typical 2.0 MW wind power turbine has its hub about 200 feet above the ground and rotor blades that are about 175 feet long.

When planning a wind farm, it is important to space the turbines appropriately, because as the wind flows through a turbine, it loses some of its energy and fans out. Therefore, industry guidelines recommend placing turbines five to ten rotor diameters behind each other, and about three to five diameters across in a line at right angles to the prevailing winds. We can use these recommendations to determine the area needed for wind farms of a given capacity. For example, consider our typical 2.0 MW turbine with a 175-ft rotor (i.e., 350 ft diameter). If the turbines are spaced seven diameters behind and four diameters across, they will each need an area of about 0.12 mi², which would provide about 16 MW of rated capacity for every square mile. With improvements in design and scaling to turbines with larger capacity, 5 MW or higher, it may be possible in the future to increase the areal density fivefold to 80 MW/mi². Right now, wind farms in the United States and Europe provide power densities around 10 MW/mi².

Many of the costs associated with laying foundations and installing grid connections do not increase proportionately with the power of the turbine. For these reasons, it may be desirable to install a larger turbine instead of smaller one. On the other hand, larger turbines may experience a higher downtime because they operate only at higher wind speeds. Indeed, analysis has shown that in many instances the net electrical energy produced over a year may be greater for a smaller turbine. Large turbines experience a greater stress from differential wind speeds along the length of the rotor and therefore require more complex engineering that may raise the scale-up price beyond economic feasibility. Finally, sometimes it is mundane considerations such as the availability of adequate roads and other infrastructural aspects that limit the size of turbines that can be installed.

Modern turbines also have sophisticated controls to alter the pitch of rotor blades and the yaw of the turbine, to best match the prevailing wind speed and direction. There are also controls for downgrading power production during wind surges, to help maintain grid stability.

Site Selection

Wind speed is naturally a very important factor in site selection: the power extracted by a given turbine varies as the *cube* of the wind speed. Thus, doubling the wind speed increases the power output eightfold. While maps such as the one in exhibit 7.2 can serve as guides for choosing a region, they are not sufficient for selecting a specific site for turbines. Local topography has a profound influence on the wind, and it is necessary to take actual measurements at the site selected for each turbine before installing that turbine. Depending on their speed, winds have been grouped into seven classes. Exhibit 7.3 gives the speed ranges of different classes of wind and the power output per square foot of the area swept by a turbine.

Specific areas that have particularly good wind energy potential include the Great Plains of the United States, the Steppes of Siberia, the southernmost coastal regions of Argentina, and islands in the North Sea and its continental shores. Currently, the most active developments are in Denmark and Germany, which are operating many large wind machines both onshore and in shallow offshore waters.

In addition to these areas with unimpeded wind flow identified above, opportunities for wind power production are also present in certain specific locations, such as gaps between mountain ranges through which air is funneled, or between regions where climate creates significant temperature differentials.

Exhibit 7.3.
Wind Speed and Power by Class (converted from the original meter/second velocity data)

Class	Speed (mph)	Power (W/ft^2)
1	<12.3	<18.6
2	12.3–15.0	18.6–27.9
3	15.0–15.66	27.9–37.2
4	15.66–16.78	37.2–46.5
5	16.78–17.90	46.5–55.7
6	17.9–19.01	55.7–74.3
7	>19.01	>74.3

An example of the latter is the temperature difference between the Pacific Ocean, cooled by the Alaskan current, and the hot central valley of California.

Onshore wind turbines (i.e., several miles or more away from the sea) are generally the easiest to install and maintain. They tend to be located on ridgelines to take advantage of the acceleration of wind from the local topography. Sometimes they are located in mountainous regions where, again, the topography can funnel wind into regions between hills. In such cases, care must be exercised in selecting the site because the effects of hills vary. Hills accelerate the local wind in some areas, while in others they serve as barriers and cause the wind to die out. Another consideration to bear in mind is the weather, particularly how prone the region is to lightning strikes or to cold spells that would ice up the rotors. To mitigate the effects of freezing, turbines have to be equipped with heaters, and heaters reduce the net energy output of the unit when they are in use.

Offshore installations are advantageous compared to those onshore because the surface roughness in the seas is less, so the winds tend to be unobstructed and strong. Often these sites are close to major power consumption centers and therefore reduce the need for long distance power transmission. Off the eastern shores of the Mid-Atlantic region of the United States there is considerable interest in installing wind farms. However, offshore turbines are also the most challenging to install, particularly in deep waters. Constant exposure to salty seawater corrodes the turbine components, creating a need for frequent maintenance. (This latter concern is not relevant for installations in fresh waters, like the Great Lakes, but there are few large freshwater lakes on Earth.) Near-shore installations can be on land within the first few miles of shore, or in water within six miles of shore. The winds close to shore tend to be stronger and more persistent, because of the temperature difference between the land and the sea.

Beyond the wind speed, other concerns that factor into site selection are the aesthetics, noise, interference with boat traffic, and paths of migratory birds. These are important considerations, particularly for people who reside or work near wind turbines. It is one thing to have a single windturbine (1–5 MW) gracing the landscape—indeed, it may be even picturesque—but it is quite a different matter when a farm containing hundreds if not thousands of such structures is needed for large-scale power production (GW range). On-shore large wind farms are also likely to be situated remotely from urban areas where the demand for power exists, and that means installation of transmission lines. There is considerable opposition from landowners and residents to building high-tension transmission lines and the associated towers. Their concerns will need to be addressed through public discourse and the permitting process. These processes take time, and that is why development of large-scale wind farms cannot be rushed.

Intermittency

Wind is fickle, and that presents a major challenge for large-scale deployment of wind energy. Seasonal variations in wind speed are large, but daily and hourly variations can be much larger. Annualized capacity factors for wind turbines tend to vary between 25% and 35%. In other words, the electrical energy generated by a 1 GW-wind farm is about a third of that generated by a coal-fired facility or a nuclear plant of equivalent power rating, which typically operates with a capacity factor of around 85%. For an electric utility that wants to use wind power, this intermittency and unpredictability mean installing an equivalent capacity of backup generators. Gas-fired turbines are generally used for backup because they have lower capital costs. The backup generators are maintained at an idling speed, so they can be ramped up to full power within a few minutes if wind ceases to blow during a time of high power demand. For example, west Texas is a region blessed with large wind potential, but the wind tends to blow at night when the grid power is already plentiful. The electricity demand is the greatest during the windless summer afternoons, when wind power there provides no relief. To fully utilize wind potential, the producers and users of wind power must add some type of electrical storage system.

In the upper Midwest of the United States, another region with good wind potential, large swaths of land have category 3, 4, and even 5 winds. Some analysts have claimed that if all of the available wind energy in this region were converted to electrical power, this region could supply all the electricity needs of the United States. However, this may not be a practical solution because the problems of producing adequate power from a variable energy source *at all times of need* are far from trivial. California has its highest electricity demand during summer afternoons, and its wind power production fits very well with its profile of electric power demands, which is driven by the high use of air conditioning and water pumping for crop irrigation during the daytime. Despite this good match, the current total annual wind production is less than 2% of the annual electricity needs of that state. Mismatches between the time of availability and demand have limited the growth of wind energy.

Like several other income sources, wind energy desperately requires the development of efficient, low-cost electrical energy storage systems.

Economics and Growth of Wind Power

In view of the substantial global potential of wind to provide carbon-free affordable energy at reasonably large scales, many countries and companies have started to build large wind farms. The growth in installed wind energy capacity has been about 30% for each of the last five years. In 2006, the total installed capacity of wind power was more than 74 GW, of which 48 GW is located in Europe and about 11 GW in the United States. Among the European countries, Germany has the largest wind power installed capacity (20 GW), followed by Spain (11 GW). Denmark, which has very favorable winds for power, has installed more than 3.1 GW of wind power and now produces 20% of its electricity from this source (the largest percentage in the world). In 2006, Denmark's wind power production amounted to about 9 TWh, or 5.8×10^{-4}

CMO. The national policies in Europe have been very favorable for the development of wind and other income resources, particularly because utilities are obligated to buy renewable power from producers at significantly higher rates (feed-in tariffs) and because carbon credits offset some of the cost of producing energy from sources not emitting CO_2.

The cost of wind power has declined substantially over the last 20 years. In the 1980s it was around 30¢/kWh, and now it is about 5¢/kWh even without subsidies—a cost comparable to the roughly 3–5¢/kWh cost for electricity from coal. Because there are no fuel costs associated with wind, the major cost factors are financing and maintenance. Installation costs are currently about $1,000/kW, and operation and maintenance expenses are generally less than 0.75¢/kWh. Other costs associated with producing wind power include systems for conditioning the power and connection to the grid. The former requires building an electrical buffer to avoid sudden surges or declines in voltage, as these would adversely affect the operation of equipment at the users' end.

The net cost of electricity from wind is markedly affected by the terms of financing and long-term guarantees for off-loading the power to the grid, since currently there is no system in place for efficiently transferring large amounts of wind power to the electric grid. This has been a major impediment for small producers, who often did not have sufficient resources to build transmission lines to connect their power source to the grid. In part driven by the state requirement to include 20% renewable energy in its portfolio, Southern California Edison embarked on a program to build and pay for the transmission lines and upgrades necessary to offload power from small wind farms, and recover the cost through the rates paid by the customers. In the 20 years that this program has been in effect, they have been able to increase the share of electricity from income resources other than large hydro to 16%. (This 16% is mostly from geothermal [10%] and wind [3%], with the remaining 3% from biomass, solar, and small hydro.)

The economics of small wind systems to serve remote rural areas not connected to an electrical grid are markedly different. In many of these areas, the per capita energy consumption is low, perhaps 150 GO/yr. Here, a 0.5 MW wind system could easily support the needs of a village of 1,000. People living here (mostly women) spend many hours each day fetching water. In such places, small wind power systems not only avert the cost of additional power transmission lines, their contribution to improving the lives of people can be substantial. By using wind energy to pump water for local use, the problem of energy storage is also minimized. Thus, while such systems may not scale to the CMO level, they can be an effective bridge technology before cost-effective large-scale systems are developed.

Producing 1 CMO from Wind Power

If we assume 30% as an average capacity factor for wind, then we will need a generation capacity of nearly 6 TW (e.g., by having three million of 2-MW turbines) to generate 1 CMO/yr of electric power. At $10\,MW/mi^2$, we can calculate the area needed for installing 6 TW of wind capacity. It comes out to more than half million square miles, or 1% of Earth's land area. To provide some perspective, this area is equivalent to one-fifth the area of the United States, or about a third that of the European Union. At $1,000/kW, the total investments for installing wind capacity to produce 1 CMO of energy each year would be $6 trillion.

Worldwide, the total energy associated with wind has been estimated to be around 1,000 CMO/yr, out of which about 40 CMO/yr may be recoverable. However, we use only something like 0.004 CMO/yr. The discrepancy between what is available and what we are using is staggering. Neither manufacturing three million turbines nor the total investment of $6 trillion over the next several decades is as daunting a challenge as is producing 1 CMO from some of the other energy sources. The challenge for scaling wind power to 1 CMO level will be in overcoming land use concerns.

Case Study: T. Boone Pickens and the Future of U.S. Wind Power

With the sharp rises in the cost of oil in 2008 and ever increasing concerns about greenhouse gas emissions from the use of fossil fuels, help for the green power industry has poured in from a variety of investors. In July 2008, former Vice President Al Gore, as a partner with the Green Team of the venture capital firm of Kleiner, Perkins, Caufield, and Byers, called for the United States to replace its entire coal-fired power with wind and solar. A few weeks prior to that announcement, oil and gas tycoon T. Boone Pickens began promoting wind power to replace imported oil and stop the transfer of more than $700 billion each year to countries that are often hostile to U.S. interests. Pickens's plan calls for building an array of 3-MW windturbines. His plan also called for building the required transmission lines from wind farms to regions where the energy would be needed. He asserted that with a one-time expenditure of $1.2 trillion, the United States would be able to produce a fifth of its electricity from wind power, the amount that is currently produced by natural gas. The surplus natural gas could be made available for use in modified cars, which Pickens asserted would therefore reduce oil imports.

We note that the amount of natural gas used for electricity is only about 0.05 CMO, while the imports of oil are about 0.14 CMO. Thus, the above plan might reduce U.S. oil imports by up to a third. In view of the highly constrained oil market, a one-third reduction in U.S. imports would offer considerable relief and could reduce oil prices. For that reason, it is worth considering the plan in a bit more detail.

According to the U.S. Energy Information Agency, U.S. production of electricity in 2006 was 4 trillion kWh. We estimate that generating for example 20% of that amount would require about 116,000 3-MW wind turbines, for a total installed capacity of about 350 GW. At $10\,MW/mi^2$, installation of 350 GW would require about 35,000 mi^2, or roughly a 15- to 20-mile-wide swath of land running the entire distance stretching north from the Texas panhandle to the Canadian border, a distance of approximately 2,000 miles. What would be the environmental impact of allocating such an area for wind farms? Even though this corridor is sparsely populated, it will likely be in the way of the flight paths of migratory birds—including some endangered species. What will our response be in that case? What about the people, mostly farmers and ranchers, in that part of the country?

As we have stated, wind power is intermittent and requires backup. In the United States, for every megawatt of wind power, utilities typically maintain an equivalent capacity in backup power to ensure a constant energy supply. The backup is most commonly provided by natural gas. Based on the experience in Denmark and other Scandinavian countries, the International Energy Agency now estimates that 60% backup should suffice. Either way, implementation of the Pickens plan will require a very substantial increase in the installed capacity of gas-fired power generators.

Finally, though, if the objective is to reduce oil imports there will have to be natural gas vehicles that can use the "freed" natural gas. If we use the Honda Civic GX as an example, we would need more than 200 million of these natural gas vehicles, each driven 10,000 miles, annually to consume the 0.05 CMO of natural gas freed by the plan. For reference, the current fleet in the United States is less than 200 million cars, so it would not be feasible to offset oil imports to any significant effect by using the freed gas in automobiles. Perhaps using it in gas-powered buses and vans would be more effective.

A one-time expenditure of $1.2 trillion to gain partial self-sufficiency in energy may sound like a bargain, but that is only part of the story. We must also consider the unstated and undoubtedly expensive need for backup power, transmission lines, and a fleet of natural gas vehicles about the size of the current U.S. fleet of automobiles along with an infrastructure to use the natural gas in vehicles, because these items are also required before Pickens's plan could have an impact on reducing our oil imports and the professed objective of stopping the transfer of $700 billion to foreign nations.[8]

Direct Solar Energy

Solar energy inputs have widespread effects. They provide direct heat and electrical energy (e.g., through photovoltaics), determine the growth of biomass, and are responsible for both the hydrological cycle and the motion of wind. Solar inputs also influence the energy potentially available from ocean currents,

8. The total volume of oil imported by the United States in 2008 was 4 billion barrels. Only at the peak price of $147/barrel would the cost of imports amount to $700 billion; at the average price of $91.50/barrel for 2008, the United States spent $430 billion in oil imports.

oceanic winds, and the thermal energy contained within ocean waters. In this section, we limit the discussion to utilization of direct solar radiation.

We mentioned earlier that the total energy falling on the 197 million mi^2 of Earth's surface in a year amounts to 23,000 CMO. This amount is equivalent to an average of 1.1×10^{-4} CMO/yr for each square mile, or roughly 230 W/m^2. Most data on insolation (i.e., rate of incoming sunlight at Earth's surface) are reported in these units of watts per square meter (W/m^2). The availability of direct solar radiation at ground level, or insolation, varies with the latitude of the location, time of year, and weather (cloud cover). A general picture of the global availability of solar energy is shown in exhibit 7.4, taken from the Texas A&M University's Web site.[9] The image shows the annualized average insolation over the entire globe. The values range from about 280 W/m^2 near the equator to less than 30 W/m^2 near the poles.

Cloud cover reduces solar availability. Hence, arid regions without clouds, such as the Saharan and Gobi deserts and the outback of Australia, receive more solar energy than do the wetter neighboring areas of Central Africa or New Guinea (although in the dry regions dust clouds do reduce the ground level availability of direct solar energy). The dry and lower latitude regions of the Southwestern United States and the adjacent Sonoran Desert region of Mexico are also blessed with favorable conditions, while the northeastern and extreme northwestern regions of the United States have less direct solar energy. Even so, the differences are within a factor of 2, not factors of 10. The average electrical generation potential in sunny, Phoenix, Arizona is 6.6 kWh/m^2 per day, while that in famously cloudy Seattle, Washington is 3.7 kWh/m^2.

The insolation data shown in exhibit 7.4 refer to the annual amounts of energy falling on a horizontal flat plate collector on the ground. From season to season within a year, there are substantial variations in the amount of incident energy. The effective incident energy also varies with the manner in which the collector plate is oriented. The amount of energy incident on a fixed plate is highest when the plate is tilted at the latitude angle along the north-south axis, so that the plate faces the sun. If power generation is most desired in the summer months, it may be beneficial to tilt the collector at an angle of 15° less than the latitude angle. For maximal power generation in the winter months, the solar collector should be tilted at about 15° greater than the latitude angle. Marked increases in the total energy harvested can be realized by incorporating a tracking system along the north-south axis, although this adds complexity and cost. Additional, smaller gains are possible by installing a two-axis tracking system, which is still more complex and costly. An example of the magnitude of seasonal variation and the impact of different collector orientations is shown in exhibit 7.5, using data for San Francisco.

9. Robert H. Stewart. "The oceanic heat budget," in *Introduction to Physical Oceanography*, Department of Oceanography, Texas A & M University, 2008, pp. 51–73. http://ocean-world.tamu.edu/resources/ocng_textbook/chapter05. Accessed August 2008.

Net surface solar radiation

Annual mean
(W/m²)

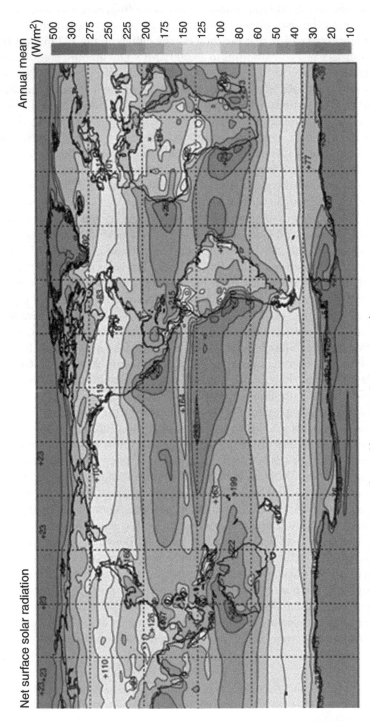

Exhibit 7.4. Annualized Global Average Insolation (from R. H. Stewart, note 9)

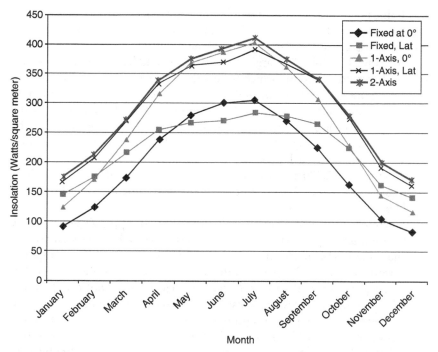

Exhibit 7.5. Seasonal Variation of Insolation at San Francisco

The monthly insolation for a horizontal flat plate collector in San Francisco ranges from a low of 83 W/m² in December to a high of 304 W/m² in July. The annualized average of 197 W/m² is close to the global average of 230 W/m², but the monthly average can vary by as much as a factor of 3. We can also see that by changing the fixed tilt angle of the collector to equal the latitude, the amount of incident radiation would increase for the winter months but decrease somewhat for the summer months. The annualized average in this case turns out to be 225 W/m², providing a gain of about 11%. The remaining three lines for the tracking systems are quite close to each other. The one-axis tracking with the tilt at the latitude angle appears to capture slightly less energy in the peak summer months of May through August than the one-axis tracking with zero tilt, but it more than compensates for that during the rest of the year. Thus, we see that gains of 25–30% can be realized with the use of tracking systems.

For many simple applications such as households, individual remote sites, and villages, the fixed flat plate is likely to be used because it will have the lowest requirements for installation and maintenance—needing only to be periodically cleaned of dust and debris. For other uses such as commercial buildings or small utility installations that are not regularly staffed, a one-axis tracker is the likely choice. The two-axis tracking systems will be used in only a few locations, such as utility stations or substations with large energy outputs and trained staff available. Staff will be needed to make repairs to the tracking system, clean the collector surfaces, and maintain the necessary electrical systems that are required for all photovoltaic units.

As shown in exhibit 7.4, large swaths of open land have average insolation values of around $150\,W/m^2$, which means that the average energy incident in one day on one square meter of land is $3.6\,kWh$, and in one year is $1,314\,kWh$. If we assume an efficiency of 20% for capturing that incident energy, the amount of electrical energy produced from that square meter would be $263\,kWh$.[10] Using this figure, we can calculate the area required to produce 1 CMO (or roughly 15 trillion kWh) of energy, and that turns out to be about 58 billion m^2, or $22,000\,mi^2$. If we look at a place like the Sahara desert, where the insolation is about $250\,W/m^2$, the area required for the collectors alone would be around $13,000\,mi^2$.

Of course, if we focused on the Saharan region to supply all global energy needs with solar, we would need to solve at least two problems. First, we would be generating power in a location far removed from centers of consumption, which would entail building a large infrastructure for storage and transport of the electrical power. As discussed in chapter 3, electricity is not readily transported over long distances and must be converted into a transportable fuel. This conversion requires consumption of energy. The second problem is that the areas needed for CMO-scale production with solar are substantial. It is likely that such an endeavor would affect the fragile ecology and biodiversity of the desert. We would want to ensure that we do not end up causing more problems than we rectify.

As we have seen, there is a great deal of solar energy available to us. There are three main ways in which we use solar energy directly: solar thermal, concentrating solar power (CSP), and photovoltaics.

Solar Thermal

Solar thermal energy, or low temperature solar, is the simplest and probably the largest source of energy that we use *if* we include noncommercial energy along with the 3.0 CMO of commercial energy. It includes using the sun's energy to keep homes comfortably warm and ventilated, to dry materials such as bricks and straw, and for water desalination. Using the sun's energy for passive heating, lighting, and ventilation is a very practical way to minimize the requirements for oil-derived energy for buildings and can be accomplished in a cost-effective manner by appropriate planning for location and building design. After the building has been constructed, the opportunities for introducing passive solar systems are mostly lost. As mentioned in chapter 3, globally almost 40% of our primary energy use is for residential and commercial

---

10. The efficiency of commercial photovoltaic systems is around 15%, and plants typically convert about 0.25% of the solar energy into biomass.

applications, which is mostly for heating, lighting, and air conditioning, and this sector therefore offers a large potential for conservation and efficiency gains.

The commercial market for direct solar heating has experienced remarkable growth in recent years, most notably in solar water heaters. Direct solar heating is most effective for low-temperature (<150°F, or 65°C) applications and is well suited for domestic use. Solar water heaters were introduced as early as 1890, but more reliable and relatively inexpensive gas-fired water heaters replaced them. The resurgence in solar water heating occurred in the 1970s following sharp rises in natural gas prices. Global installed capacity of solar heating is more than 100 GW and is increasing at the rate of about 15 GW/yr. Most of this increase (~10 GW/yr) is occurring in China, where gas and electricity are less affordable and often less reliable than the sun. Rapid increases have also occurred in Germany, Israel, and several Mediterranean countries, where governments have adopted prosolar policies.

The technology for solar heaters is quite straightforward. It consists of flat plate or evacuated tube collectors, a circulating fluid, and a holding tank for the hot water. A flat panel collector consists of a metal plate, often blackened to enhance absorption of solar radiation, and a set of pipes through which water (often containing an antifreeze) is flowed to transfer the heat to a hot water tank via another heat exchanger. The panel and the tubes are oriented to maximize solar exposure. A simpler version of this device, used for heating swimming pools mainly in Australia and the United States, consists of a blackened plastic pipe through which water is circulated. Since there is no insulation, the device is useful only for applications where the operating temperature is not substantially above ambient. As the fluid heats up it also gives off energy to its surroundings, and thus becomes less efficient at higher temperatures. Placing the plate and tube assembly in an insulated box with a transparent glass cover reduces this heat loss, and that configuration is used in many flat-plate solar collectors. These devices work well in regions with plenty of sunshine. Where the temperature drops below freezing, evacuated tube collectors are preferable, which consist of blackened absorbing pipes placed inside an evacuated glass tube. The vacuum offers a better insulation because it eliminates convective heat loss from the absorbing tubes. These devices are also somewhat more effective under gray skies.

Solar water heater units currently cost about $300 in emerging economies such as India and China, and less than $2,000 in the United States. At that price, many homeowners can afford to have both a water heater plus a gas water heater as a backup. Widespread adoption of solar thermal systems has a low economic barrier, well suited for places like China and India where we are likely to see the greatest growth in energy use in the coming decades. For these reasons, we expect that solar heating will contribute significantly in the future, with its greatest contribution being to reduce the need for other power sources. Indeed, solar thermal is the unsung hero of green energy.

Concentrating Solar Power

The solar collectors described above can only heat fluids to temperatures of about 150°F (65°C), which is not sufficient to run a heat engine. Achieving significantly higher temperatures requires collecting sunlight over a larger area and concentrating it. There are several collector designs, including parabolic troughs, power towers, and dishes. The trough design is a long parabolic mirror with a pipe along the axis through which a liquid, generally oil or a molten salt, flows. The concentrated sunlight heats the liquid and the heat is used to generate steam that then drives a turbine to generate power. In the tower design, a large number of flat mirrors are arrayed around a central tower. They are mounted on sun trackers or heliostats so they can continually direct the sunlight to one small area, where typically the solar intensity can reach as high as 500–700 times that at Earth's surface. The heat is intense enough to melt many salts. Indeed, molten salts are the preferred fluid in these systems. The hot molten salt is used to produce steam for a power turbine, or it can simply be used to store heat for use later. The molten salt is an effective medium for storing heat, which is considerably simpler than storing electricity.

In the third design, parabolic dishes, a Stirling engine is mounted at the focus of the dish. Stirling engines are pure heat engines with a hot zone and a cold sink. There is a fixed amount of gas in the engine that acts as the working fluid transferring heat from the hot zone to the cold sink. The heated gas expands and moves a piston. As it expands the gas gets exposed to the cooler sink of the engine. The cooling gas contracts and pulls back the piston, and the cycle goes on. There is no exhaust, and no need to plumb any fuel or gas in and out of the system. The requirement is that a temperature difference be maintained between the hot zone and the sink. In parabolic dishes, the sun's energy is focused on one end of the engine, which keeps it hotter than the opposite side that is in the shadow and exposed to the ambient air. The mechanical power from this engine is used to make electricity.

While the solar-thermal power-generating units can be operated to supply base and/or intermediate power, their low capital costs make them more likely to be used as base-load units with very low fuel costs. If the large units are operated independently as solar-only facilities, they will have the same dependency on overall electricity as for photovoltaic power, discussed below.

The cost of the parabolic trough optical unit is around $50/m² and is one of the major cost elements in that system. Even if the system were located at a place of high insolation, such as 250 W/m², the contribution to capital cost of the mirrors alone would be on the order of $200 per kilowatt of capacity. A useful target for most power systems with low fuel costs is less than $2,000/kW. In the tower design, the cost of mirrors is less because they are of the flat plate variety, and the major costs are associated with the heliostats used to track the sun. While there are significant differences in how parabolic trough and power tower systems harness the sun's energy, they both use the same steam turbines for power generation as are used in gas- or coal-fired power plants. For that reason, trough and tower CSP systems are well suited for use in conjunction with existing fossil fuel systems. This combination also minimizes the investments and lowers the overall cost of producing solar electricity.

The trough and tower CSP design systems have been demonstrated at sizes of 100 MW of power output and are considered ready for utility-scale deployment. Their conversion efficiencies are up to 25%, compared to the average 33% efficiency of conventional fossil-fired power units, but the fact that there are neither fuel costs nor emissions associated with their operation more than makes up for the lost efficiency in terms of cost. The dish systems are still under development and are expected to have solar-to-electric efficiency of 30%.

Commercial viability of CSP was first demonstrated in the 1980s with the Solar Electric Generating Stations project in the Mojave Desert of southern California. A large number of units were built, and they collectively reached a capacity of more than 350 MW. However, shortly after the demonstration project, two things happened that caused the building of CSP plants to be abandoned. First, natural gas prices plummeted, and power from gas-fired plants became much more competitive; second, the Luz Corporation that had developed the trough technology became embroiled in a financial crisis that brought the development process to a halt.

With the recent rise in natural gas prices, as well as the increasing demand for carbon-free sources of energy, CSP has seen a revival. Although the cost of power from coal-fired plants (3–5 ¢/kWh) is lower than that from CSP units (0.15–20 ¢/kWh), there is great reluctance on the part of banks to finance new coal-fired power generation systems. The reticence stems from the uncertainty regarding the future cost of CO_2 emissions that could alter the economic equation. However, CSP systems built in conjunction with fossil fuels, used as supplemental heat sources, alleviate the low duty cycle of CSP-alone systems, and diminish the potential carbon tax penalties compared to system using fossil fuel power only. This synergy with other thermal power generation systems bodes well for the deployment of CSP systems.

Solel, Inc., is currently generating power from the nine plants in the Mojave Desert originally installed by the Luz Corporation and is building a 553-MW solar thermal plant using the same parabolic trough technology the Luz Corporation pioneered. This plant is scheduled to come on line in 2011. In the neighboring state of Nevada, Acciona Energy built a 64-MW power facility called Nevada Solar One, at a total expenditure of $260 million, or slightly more than $4,000/kW. BrightSource Energy has announced development of a series of solar plants in the Mojave Desert using Luz Power Towers with a combined capacity of 1.3 GW. More than 10 CSP stations are being built in the southwestern United States that together will have a power generating capacity of more than 3 GW. Many large projects are also under way all around the globe. One notable example is the 600-mirror solar tower called PS 10 in Seville, Spain, that generates 11 MW of electricity. It was developed by Solucar, a subsidiary of Abengoa, which is also planning two other 20-MW tower projects, PS 20 and AZ 20. Also in Spain, ACS Cobra and Solar Millennium are building 50-MW$_e$ parabolic trough systems, Andasol I and II, in Andalucia. A 140-MW facility is being built in Algeria, and similarly large plants are planned in India, Egypt, and Morocco. Completion of these large commercial projects will go a long way in providing benchmark data for CSP systems as well as push the industry far along the learning curve so that it may also drive down the installation costs from $4,000/kW to less than $2,000/kW, which would make the price of electricity from these plants competitive with coal plants even without any carbon tax.

Photovoltaics

The third way of using solar power is by using photovoltaic (PV) materials. Photocells, which are devices made from PV materials, were known in the 1840s. The basics of semiconductor technology on which today's PV units are based were under development more than a half-century ago. Silicon-based devices, the most common of PV devices, entered the field in 1941 when researchers at Bell Laboratories developed techniques creating positive and negative sites (p-n junctions) in a single crystal of silicon. The importance of the p-n junction will become evident when we discuss the basic technology below.

Subsequent advancements in practice and theory of the electronic behavior of material in the solid state led to practical developments in PV power, also unleashing the enormous semiconductor industry of microprocessors and memory chips. Development of today's PV technologies was spurred by the need for power onboard spacecraft in the 1960s. Photovoltaic power came into wide use in the 1970s, for very small power applications such as pocket calculators and watches, and in watt- to kilowatt-sized remote applications, including buoys, communication relays, and various other off-grid applications existing at least a mile from electricity distribution lines.[11]

The uses of PV devices have been growing steadily and now include limited power production for villages in developing nations, providing electricity for both community and individual needs.[12] In these applications, batteries are used for storage to provide lighting and other nighttime uses. Photovoltaic energy is used in developed nations to power remote sites, and increasingly as an adjunct to grid-supplied power at homes and offices.

11. In many remote applications, including general village needs, photovoltaic systems are replacing existing small gasoline or diesel-electric power generators.

12. These uses include refrigeration for medical supplies, radio and television reception for schools and homes, and cooling fans and lighting for a few hours at night. PV units are also used to power small irrigation pumps. Some of these units are small enough to be mounted on a wheelbarrow that can be moved from field to field. Most of these applications fall into the noncommercial energy category.

In the production of PV energy, light striking a semiconductor material excites some of the electrons and renders them mobile. The conversion device has no moving parts and seems to offer a very simple way of producing electricity. It consists of a semiconductor material encased in a container with a transparent front cover and a light-reflecting rear surface. The electron excited by light drifts to the positive end of the p-n junction, and electrical contacts there lead to an external circuit where power generated by the device can be used.

While photocells and semiconductor devices have much in common, they have one distinct difference: photocells are limited by how deep the light penetrates into the material, while most semiconductor applications can utilize the full depth of the material. Another difference is the very high purity requirements for semiconductor devices as opposed to photocells—a requirement that affects the economic viability of large-scale PV power. Photovoltaic devices have a materials cost advantage in that they can use material rejected by the semiconductor industry.

Finally, the market demand for semiconductor devices has resulted in mass production opportunities and commercial money for further development. For many years, the PV industry was dependent on government funding driven by space-age needs. Space has ample, continuous sunlight—a luxury not available for Earth-bound power applications—and the overall costs of on-board space systems made the otherwise extremely high costs of PV power trivial in that application.

A PV system for power generation avoids some of the shortcomings of fossil fuels, insofar as extracting power from fossil fuels requires several energy conversion processes. First the fuels are burned to heat a fluid, then this fluid is passed through a turbine to produce mechanical motion, and last the mechanical energy is converted to electrical power. In contrast, in a PV system light energy is directly converted to electrical power. In principle, PV conversion efficiencies could be higher and operations considerably simplified. Although operation of PV systems is already quite straightforward, for various theoretical and practical reasons the realized efficiencies remain limited. In the following section we discuss basic PV technology to gain a better understanding of its limitations, and what is being done to overcome them.

Basic PV Technology

Silicon is one of a class of materials known as semiconductors.[13] Ordinarily, silicon is an insulator as opposed to a conductor, and all of its electrons are bonded to atoms in what is called the valence band. If somehow one or more

---

13. In this section we discuss photovoltaic phenomena related to silicon, and the principles apply also to other semiconductor materials used to produce photoelectric power. Many semiconductor materials alone, or in combination, can be used to produce power. These include amorphous as well as crystalline silicon, germanium, cadmium sulfide, cadmium telluride, indium phosphide, and gallium arsenide.

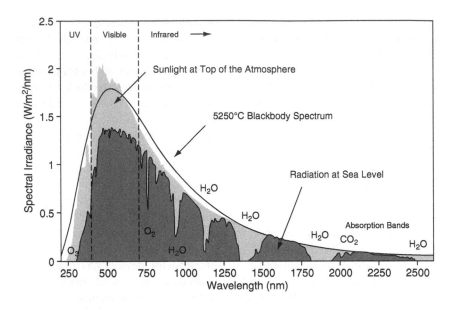

Exhibit 7.6. Solar spectrum arriving at the Earth (from Wikipedia)

of these electrons is excited to the conduction band, silicon can be made to conduct electricity. The energy gap between the valence and the conduction bands of semiconductors is referred to as the band gap, and it largely governs the voltage of the PV cell. For crystalline silicon, the band gap is 1.08 eV (electron volts), which means that absorption of photons with energy in excess of 1.08 eV will result in the excitation of electrons to the conduction band. A photon of light with energy of 1.08 eV corresponds to a wavelength of 1,500 nanometers, which is in the near infrared region of the spectrum. Sunlight photons with higher energies (e.g., those in the ultraviolet and visible portions of the spectrum) will also excite electrons of silicon to the conduction band. Beyond the energy needed for excitation, any excess energy that they have is soon degraded to heat. This energy loss is one of the principal causes limiting the efficiencies of solar cells.

The solar spectrum in outer space and at sea level is shown in exhibit 7.6. The distribution of the intensity of photons of different wavelengths in outer space corresponds to that emanating from a black body at 5,250°C. As the radiation passes through Earth's atmosphere, many portions of the spectrum are absorbed. Less than 5% of the energy of the sunlight arriving at Earth's surface is in the ultraviolet region (shorter than 400 nm). The visible portion of the spectrum (400–800 nm) contains about 60% of the energy, and the remainder is in the infrared region. Photons of wavelengths longer than 1,500 nm comprise no more than 10% of incident solar radiation and essentially pass through silicon. All photons with wavelengths shorter than 1,500 nm will have energy sufficient to excite the valence electrons in silicon to the conduction band.

Excitation of the silicon electrons is required but is not enough to produce power. The excited electrons must be persuaded to carry a current, which can occur if a p-n junction is present. This junction is created by the juxtaposition of two pieces of silicon that have different impurities—commonly boron and phosphorus. These impurities, or dopants, differ in their electronic structure from silicon. Silicon has four electrons in its outer shell, boron has three, and phosphorus has five. Incorporating boron in a silicon structure creates a region that is essentially one electron short, and this acts as a trap (hole) for electrons. Such dopings with impurities having fewer electrons are also referred to as positive- or p-doping. In contrast, doping with phosphorus creates a silicon structure that essentially has one electron in excess, resulting in a negative- or n-doping. Having one face of silicon wafer p-doped and the other n-doped creates a p-n junction. The p-n junction induces an electrical gradient such that if connected by an external circuit, the electrons will have a tendency to move from the negative n-junction towards the positive p-junction. This tendency would be countered by the buildup of charges, but the light used to excite the electrons within the silicon causes the electrons to move toward the phosphorus-doped n-junction, and then through the external circuit. As each electron moves in to fill a hole, it leaves behind a (new) hole. As a result, within the device electrons are moving in one direction and holes are moving in the opposite direction. The resultant current flow and the voltage differential (band gap) create useful power.

Preparation of silicon-based photocells often begins with an ingot of p-doped polycrystalline silicon. A batch of silicon is melted along with the boron dopant, placed in contact with a mother crystal, and then slowly cooled to produce cylindrical ingots of polycrystalline silicon with a diameter of perhaps 6–12 inches, and length of 3 feet. These are the same ingots that go into the manufacture of computer chips, except that the purity requirements for solar cell applications are less stringent than for computer chips. The ingot is then wire-sawn to produce thin sheets (0.01 inch thick). Material losses in this step are high. Further loss of material is incurred when the circular wafers are squared-off so they can be packed in rectangular arrays, also known as modules, without leaving a lot of open space between them. The sawed off material can be recycled through the crystal growing process. The top surface of the wafer is then doped with phosphorus either by a diffusion process or by ion implantation, to make a p-n junction. Thin metal wires are laid down to collect the electrons. The other side of the wafer is then backed by a layer of a conducting metal, commonly aluminum. The entire cell is protected from the environment behind a glass sheet, and many cells are assembled and integrated to form modules.

With this background, we can now discuss some of the technical challenges in the design of PV cells. One challenge relates to the thickness of the wafer and the efficiency of a PV cell. For maximum efficiency, the electrodes must collect all the electrons and holes generated by the photons. However, in many instances, the electrons and holes recombine before reaching their respective electrodes. This recombination can occur at defects in the crystal. To minimize such losses, it is necessary to have high-quality crystals that are preferably very thin, so that the distance that the electrons and holes have to migrate is kept to a minimum; however, thin wafers do not have good mechanical properties. Thicker wafers also ensure that more incident light gets absorbed by the silicon, since the amount of light absorbed depends on the thickness of the wafer. These countervailing trends require striking an optimal compromise.

Another factor that lowers the efficiency is the mismatch between the energy of the photons and the band gap. As mentioned above any photons of energies less than the band gap cannot lead to an electron-hole pair. Furthermore, although the photons with sufficient energies are converted into electron-hole pairs, any excess energy that the photon had over that required to cross the band gap gets dissipated as heat; in other words, it is essentially wasted.

From the band gap and the known solar spectrum, we can calculate the theoretical efficiency of any semiconductor material. This number represents the fraction of light energy that was used to produce an electron-hole pair in the first place. For silicon the theoretical maximum efficiency is 25%. Good manufacturing practice can produce cells with efficiencies of 15–17%. The band gap of gallium arsenide, another semiconductor used for solar cells, is 1.4 eV. It is ideally suited to the solar spectrum. Gallium arsenide uses a smaller portion of the solar spectrum, but its higher band gap results in a higher overall voltage for the device. The net result is a theoretical efficiency of 25.1, which is essentially the same as crystalline silicon. The main advantage of gallium arsenide is realized in space, where the high-energy end of the solar spectrum is not attenuated by Earth's atmosphere. In space, the theoretical efficiency of gallium arsenide is almost 30%. It is the preferred material for providing power to satellites and rovers on Mars, because achieving higher efficiencies translates to using less material and less weight. Amorphous silicon, a material that can be produced inexpensively by a screen-printing process, has an even higher band gap of 1.7 eV, but its efficiencies are generally much lower, around 6.6%. This is because the lack of crystallinity leads to too many losses from electron-hole recombinations. As is often the case, there is a trade-off between the cost of a material and the efficiency it can deliver.

Recent Advances

There have been a number of advances directed at improving the efficiency of solar cells. We present here examples of a few of these innovations. We chose these examples not because we are endorsing these technologies or believe that these are the most promising approaches for successful commercialization, but rather as representative of the variety of approaches researchers are investigating.

To overcome the dilemma of wafer thickness, SunPower has designed a cell that has the p-n doping in alternate lines on the back surface. The bulk of the wafer is pure silicon, so it presents minimal opportunities for the wasteful recombination of the electron-hole pair. The top surface of the wafer is textured and coated with an antireflective coating, to minimize reflection of incoming light and to trap light within the wafer so it will bounce back and forth many times before escaping. The net result is an increase in the length that light travels in the wafer, and thus a more effective absorption of the light even if the physical thickness is about a third of the normal wafers. With these improvements, SunPower reports cell efficiencies of greater than 20%. They are installing a 2.5-MW tracking system as part of a 15-MW PV project at Nellis Air Force Base in Nevada.

Cadmium telluride (CdTe) is a semiconductor material with a band gap of 1.56 eV, which makes it suitable for PV applications. Thin films of this material can be vapor deposited on glass substrates to fabricate PV panels that are less expensive than Si wafers and more efficient than amorphous silicon. In 2007 First Solar reported reducing the cost of CdTe PV panels to $1.00/watt. Another company, Bloosolar, has developed a vapor deposition process for growing arrays of thin rods of CdTe on a substrate. The rods are a few nanometers in diameter and a couple of hundred nanometers long, and the arrays look like a brush under a microscope. The rods are then coated with a transparent electrically conducting material. This configuration also has a short distance over which the electron-hole pair has to migrate (nanorod diameter), while still allowing ample opportunities for the photon to be absorbed by the PV active material. A photon that is not absorbed by the CdTe in the first attempt gets many more chances to bounce between bristles. The downside of CdTe panels is the toxicity of the cadmium, as it requires special handling and disposal practices.

The designs discussed above are typical in their reliance on a single material with a single band gap, also known as a single junction cell. Advances have also been made in designing cells that have multiple materials, each with progressively lower band gap. The top layer with the highest band gap absorbs light of shorter wavelengths and allows the longer wavelength light to pass through this top layer. This transmitted light is then absorbed by subsequent layers. In this manner more of the energy of the light is converted into electrical power. Cells with efficiencies as high as 38% have been reported. Designing cells with multiple junctions is more complex because it requires managing a larger number of materials and careful matching of the current produced in each layer.

One promising group of PV materials is a combination of copper, indium, gallium and selenium, commonly known as CIGS. As a direct band-gap semiconductor, CIGS can generate far more electricity from a given amount of material than a typical silicon-based single-junction PV cell. A CIGS film as thin as 1 micron produces a photoelectric effect equal to that of a crystalline silicon wafer 200–300 microns thick. By using less than 1% of the semiconductor material required by crystalline silicon cells, CIGS devices offer sustainable cost advantage. Researchers at the DOE's National Renewable Energy Laboratory have reported CIGS photocells with efficiencies of 19.5%, which is the highest for a thin film technology. Miasole, a California company, is developing a roll-to-roll printing process for mass-producing CIGS cells.

Another approach to using less PV material is to use lenses to concentrate the light and focus it on to the PV material. Fresnel lenses, which can be stamped out inexpensively, are generally used for this purpose. It is important to remember that the areal requirement of a PV system is not affected by the use of lenses; only the amount of the active material that is used is reduced. This factor though can be very significant, particularly for some of the more exotic multiple junctions devices. Thus Spectrolab, a subsidiary of Boeing, recently announced its three-junction solar cells with efficiencies of 40%. The design uses light concentrated 100 fold and also uses what are known as metamorphic crystals. In metamorphic materials the lattice mismatches between layers of different compositions are minimized, thereby reducing the opportunities for wasteful electron-hole recombination.

In 2005, the U.S. Defense Advanced Research Procurement Agency (DARPA) announced a grand challenge for producing very-high-efficiency solar cells at low cost. The research team, led by the University of Delaware, included industrial partners such as Corning, Emcore, BP Solar, DuPont, and Blue Square Energy, as well as other institutions, including the Massachusetts Institute of Technology, the University of California–Santa Barbara, the University of Rochester, and the National Renewable Energy Laboratory. According to DARPA, the team has leveraged advances in several key areas, such as biofabrication, advanced solar cell architecture, and nonimaging optics, and is poised to deliver a roll-to-roll process for mass-producing multijunction solar cells with efficiencies pushing 51%. This advance doubles the efficiency possible with silicon and is a huge step forward!

Global PV Market

The PV market has seen a phenomenal annual growth of between 25% and 50% since 1990. The growth was initially driven by subsidies offered by the Japanese government. According to the Japanese government agency NEDO, in 1994 the installation cost in Japan was more than 1,900 yen/W, and the incentive offered by the government was 950 yen/W, or 50% of the cost. This program of incentive was designed to tail off to zero by 2005, and the cost of installations decreased steadily during this period and was around 700 yen/W in 2005. The resulting rapid increase in the demand for silicon led to shortages and some downturn in 2004, but additional silicon production capacity and the use of alternate materials soon alleviated those shortages.

Prior to 2005, most of the silicon was used for making computer chips, and the PV market was served by the discards of the chip-making processes. Since 2005, however, PV applications now command the larger portion of the silicon market. Germany began its aggressive growth in the PV arena in

2005 by mandating a feed-in tariff of 0.55 euro/kWh for PV power. More than half of the annual global increase occurred in Germany alone in 2005, where 600 MW was added in one year and grid-connected solar PV in Germany surpassed that of Japan. In 2006, Germany installed more than 1 GW of PV systems. Global cumulative PV capacity was projected to be 12.4 GW at the end of 2007, more than 90% of this used for grid-connected systems. More than 75% of the global installed capacity is in Japan and Germany. Japanese manufacturers Sharp, Sanyo, and Kyocera and Germany's Q-Cells are among the world's top producers of PV systems.

Major installations include the 10-MW Solarpark in Bavaria, Germany, covering 62 acres in three locations. It consists of 57,600 panels mounted on trackers and has been in operation since the end of 2005. In Beneixama, Spain, a 20-MW PV facility covering 124 acres recently came into service, and in Australia, a 154-MW facility is being planned for completion in 2013. In all these cases, there is a strong interest on the part of the national government to promote industries in the "clean energy" arena, to meet the dual objectives of helping achieve the Kyoto protocol mandates and providing economic stimulus for a new growing industry. In the private sector, the large solar PV installation at Google headquarters in Mountain View, California, is noteworthy. Built on the roofs of almost all its structures and parking lots, this installation of 4.5 acres is rated at 1.6 MW and is expected to save the company almost $400,000 a year.

Cost reductions in manufacturing methods are equally, if not more, important than improving cell and module efficiencies are to the commercial success of PV power. Both less costly manufacturing techniques and lower rejection rates will assist the economics. At present, only the wafer technology using polycrystalline silicon is used for generating power for home and office use. With respect to the energy balance, these PV modules are a net energy producer after about five years. In other words, the modules have produced more energy in total after about five years of operation than was expended in processing the silicon and manufacturing of the module.

The cost of PV modules has been declining steadily since they were first introduced for space applications in 1975. According to a compilation by the Earth Policy Institute, the cost of modules in 1975 was roughly $100/W (converted to 2007 dollars). It decreased steeply to about $10/W by 1985 and has continued to decline ever since, albeit at a slower rate. In 2008, the modules cost around $3.80/W.

The installed cost is significantly higher because it also includes costs for the physical structure, an inverter, and connection to grid (or to a power storage device). In the United States the balance of system costs currently run more than $4.00/W, such that the cost of electricity from solar PV is between 30¢/kWh and 60¢/kWh. This cost is considerably higher than what most residential users pay. However, because much of the solar power is generated during peak hours, when the cost of electricity can be even higher, installation of a home PV system could be justified strictly on cost grounds, if the homeowners are paying that high peak price for power use during those periods, or if the utility would buy back power from customers with solar PV systems at those peak prices. Even so, installation of a home PV system entails a substantial upfront outlay, and the payback could take more than 15 years, which is more than half the expected life of the system. In a recent study, Severin Borenstein of the University of California–Berkeley analyzed the cost of PV installations ($86,000 for a 10 kW system) and the potential savings from use over their lifetime.[14] He then calculated the net present value of those savings, to compare them against the cost of installing the system, and concluded that installing a PV system cannot be justified solely by economic considerations. Furthermore, according to his analysis, it would require imposition of a carbon tax of more than $100 per ton of carbon to make the cost of PV power competitive with that from coal. This tax is several times greater than what is currently being considered for CO_2 abatement.

Of course, we must realize that the reasons individuals choose to install PV systems go far beyond economic considerations alone. Subsidies from the government and utilities, with a desire to reduce greenhouse gas emissions, are the primary motivators. Incentives used include relief from taxation of property, direct subsidies, and opportunities for resale of any excess power generated by individuals and businesses to utilities at the same rates that the utilities charge these customers—not at the utilities' lower cost of generation.

A justification offered for these incentives is that of "leveling the playing field" by providing assistance to a fledging industry. Such incentives were given, or are still being given, to other sources of energy.[15] As the PV power industry grows, the substantial incentives now offered may decrease, either because PV system costs finally reach economically competitive levels or because accumulated costs of the subsidies are judged unaffordable by the public. As the number of installations grows, developers argue (and hope) that with the learning experience new innovations will come along to reduce costs leading to further increases in the demand for PV systems. Through several cycles of price reductions and increased production volumes, the effect will be to reduce manufacturing costs to a point that subsidies are no longer needed.

Essentially all of the PV applications in developing nations are also subsidized by governments and/or international organizations that wish to extend

14. Severin Borenstein "Electricity Rate Structures and the Economics of Solar PV: Could Mandatory Time-of-Use Rates Undermine California's Solar Photovoltaic Subsidies?" Center for the Study of Energy Markets Working Paper #172, University of California Energy Institute, September 2007.

15. These incentives include, among many, depletion allowances for fossil fuel producers, and government research and development assistance and insurance protection against costs of major accidents offered specifically to nuclear power operators.

the benefits of electricity use, without the difficulties and expenditures associated with extending transmission and distribution services from a large electricity generation and distribution system to the very low-demand locations. The subsidies include low-cost loans based on long payment periods and low interest rates.

Future advances in multijunction cells, coupled with advances in thin film processing, hold the promise of significant reductions in the cost of PV modules, and this is key to large-scale deployment of solar PV. Perhaps equally promising are new business arrangements in this industry in which traditional roofers and builders take on the role of selling PV systems. It is much less expensive to install PV systems at the time a house roof is put up than to retrofit an existing roof. As mentioned above, the costs associated with installing the supporting structure of a PV system can be as much as 50% of the total cost. Another emerging business model is for companies to "rent" rooftop space from homeowners. The companies pay for the upfront costs of installing a PV system, thus removing the major financial obstacle for homeowners. In return, the companies take any direct subsidies and also revenue from the sale of any surplus power back to the utility.

Producing 1 CMO by PV Systems

In this section we address what it would take to generate one CMO of electric power from PV systems, how many units it would take, and what the level of required investments might be. For this analysis we first consider a commercially available 2.1-kW home PV system that consists of 10 rectangular modules, each measuring 3 ft by 4 ft. With a typical value of sunlight availability of 20%, such a system would produce 3,600 kWh of electricity over the course of a year. If we wish to produce 1 CMO of energy (15.3 trillion kWh) from these PV systems, we will need a total of 4.3 billion such systems! That's one for every two persons expected to dwell on Earth in 2050. A million roofs, which is a target expressed in a slogan by the California Energy Commission, would produce only 3.6×10^9 kWh, and would provide only about 1% of the state's annual electricity use. Rooftop systems are clearly limited in meeting large-scale energy demand.

An alternative to use of individual PV units on homes and offices for producing CMO-scale electric power is to install large solar parks. Germany has been at the forefront of installing large systems, but to date the largest of these parks have capacities in the 10 MW range. Even in solar-progressive Germany, opposition to building large systems is mounting as people realize the visual impact of having shining arrays of PV systems on rooftops of the city's historic buildings. As we discussed under the general considerations for solar systems, to generate 1 CMO of power per year from a system that converts 20% of solar energy into electricity, the total area for the collectors would need to be around 20,000 mi². This area is considerably smaller than needed for wind (half a million square miles) or biomass systems (half to several million square miles). Obviously, the challenges of finding areas that (a) have adequate sunshine, (b) are also close to regions of energy consumption, and (c) do not raise serious environmental concerns stemming from change in the use of land such as ecological disruption and impact on the habitat of endangered species or biodiversity are not going to be easily overcome. As for all energy options, we will need to make individual and comparative value judgments.

The estimated PV contribution to total energy in 2050 ranges from 2% to 10%, and cumulatively over the period from 2000 to 2050, the contribution from PV is estimated from a low of 0.33% to a high of 1.7%.[16] In 2005, the total power produced by PV energy was 0.001 CMO. For PV power to increase to 1 CMO by 2050, it would require about 12 doublings, or roughly a doubling every four years, or an annual growth of 18%. In this scenario we would expect PV to become a major supplier of power in the latter half of the 21st century, assuming, of course, that this growth rate could be sustained. While the last few years have seen annual growth rates in PV above 40%, the PV market was starting from a relatively low base. After the market has grown significantly, continued expansion is likely to run into resource limitations—of material, land area, or workforce—that would slow the growth rate.

As the use of PV systems increases, matching their power generation to consumer demand may be difficult. For example, at around noon on brightly lit summer days, the energy demand created by air conditioners could be supplied by PV equipment, instead of being filled by fossil-fuel-fired peaking generators as are used now. In contrast, if these demands come on winter days with little sunlight, PV sources would provide only an unimportant addition to the power required from other traditional resources. Some mitigation of these matching problems could be achieved through use of effective energy-storage systems. As discussed in chapter 8, implementation of a smart grid could also help relieve supply and demand matching problems.

If efficient, economical storage is unavailable, then PV power use will be limited to the times when the sun is shining and the electricity system it serves can use the energy produced. An economically practical limit for PV capacities in solar-favorable areas of the United States has been estimated to be about 20% of the utility's total installed capacity. Since some other places in the United States and the rest of the world will not have summer peak electricity demands that coincide with peak solar production, we assume

---

16. Certainly other photovoltaic applications, for example, for households not connected to an electrical distribution grid, could add to the totals above and reduce the amount of fossil fuel used to produce needed electricity, but these applications would have negligible effect on world production. Production of hydrogen through electrolysis with an 80% energy efficiency followed by regeneration of electricity in a fuel cell with an efficiency of 60% (current efficiencies) means that the original generating equipment (photovoltaic cell or wind machine) requirements are effectively doubled, or energy production rates halved.

here that the world average for usable electricity of PV origin will be instead 10% of the total.[17]

To "guesstimate" the total PV use in electricity systems (the largest portion of the total energy use) we must (a) select a total energy demand and (b) estimate the electricity fraction of total demand. In chapter 4 we considered four scenarios with total energy-demand estimates ranging from 3.2 to 9.4 CMO/yr by 2050. If we further assume that electricity use is either 40% or 50% of total energy use, the range of values for the total electricity consumption would be 1.3–4.7 CMO, and the contribution from PV would be 10% of that, or between 0.13 and 0.47 CMO. A likely scenario is that the total energy consumption is 6.0 CMO, electricity use is 50% of that, and the PV sector grows to 0.3 CMO by 2050.

Photovoltaics have the capability to provide a significant portion of our energy once the technology becomes more cost-effective, and by that we mean the cost not just of the PV panels but of the *whole* system. Beyond current applications, PV systems will have an increasingly significant role if we shift towards driving more electric or plug-in hybrid cars in the future. But because PV systems do not scale to the CMO level, they are not likely to become a major contributor to our total energy requirement for the next 50 years. With this limitation in mind, we as a society have to reexamine how much developmental support or subsidy we should provide this industry vis-à-vis others.

Biomass

Biomass comprises all materials derived from living organisms including plants, animals, microorganisms, and their wastes. The distinction between biomass and our fossil resources, coal and oil, is principally one of time. Fossil resources are formed from organisms that lived millions of years ago. In contrast, biomass generally refers to materials that were produced by organisms that were living until recently, often within the last decade. Almost all of that material is ultimately derived from photosynthesis, by which plants and certain microorganisms use the energy from sunlight when combining atmospheric CO_2 and water to form sugars and other primary building blocks. Collectively, each year terrestrial and aquatic plants and microorganisms produce about 20 CMO of biomass. Animals do not have the ability to photosynthesize, so they eat plants for sustenance. For a long time, humans used biomass (mostly wood) as the primary source of energy, then wood gave way to more convenient fuels such as peat and coal. Burning biomass releases stored energy and does produce CO_2, yet because the plant had fixed the CO_2 in the first place, growing and using biomass as a primary energy source does

---

17. For example, demand in Northern Europe and the Russian Federation is highest at the times that solar energy is least available.

not *add* to the greenhouse gases in the atmosphere. For this reason, biomass is considered as a carbon-neutral fuel, and there is much interest in finding ways to use biomass as an energy source.

While our other income sources of energy can be used to produce electric power, only biomass lends itself to production of liquid fuels, and our current transportation system depends almost exclusively on liquid fuels. As mentioned earlier, 93% of the energy used in transportation currently comes from petroleum. Thus, finding a way to derive some of that fuel from biomass is a strong motivation for countries that have limited petroleum resources. Vegetable oils from crops such as soy, rapeseed, jatropha, coconut and palm can be used for making biodiesel, and starchy crops such as corn and sugar cane can be fermented to produce ethanol or even hydrocarbons. In the following sections, we first examine the global potential for biomass and then review some of the technologies being developed and how they might be scaled. We also address the extent to which each approach helps to displace petroleum or other fossil resources and to reduce greenhouse gas emissions.

Range of Biomass Materials and Options for Recovering Energy

Biomass refers to a wide range of materials, and that opens up an even wider range of options for its processing and utilization:

1. Oil-bearing plants. Humans have used oil-seed bearing plants such as soy, corn, and palm for many centuries. They are recognized as a good source of energy in the diet. A relatively recent addition is canola: it is a cultivar of the rapeseed plant that was developed in Canada. There are also the nonedible oils from plants like jatropha, pongamia, and jojoba that are now being cultivated for the express purpose of converting their oil into biodiesel. Oils could also be produced from algae for this purpose. The chief advantage of algae is that they are very efficient at harvesting sunlight and therefore require considerably less land area compared to other crops.
2. Starch and sugary plants. Crops such as sugar cane, corn, cassava, wheat, barley, and sugar beets are rich in starches and sugars, and they can be used for producing fuels like ethanol and butanol by fermentation. The sugars are fermented with yeast, and the starches require a relatively simple step of saccharification before yeast can act on them.
3. Lignocellulosic biomass. This biomass consists of woody trees, grasses, and crop residues like rice hulls and corn stover. Trees such as poplars and grasses such as switchgrass and miscanthus grow very quickly and can be cultivated as energy crops. Cellulosic biomass can also be fermented into fuels; however, it requires a few extra steps to remove the lignin and convert the cellulose and hemicellulose that are essential parts of most biomass into fermentable sugars. Lignocellulosic products may also be treated in a thermochemical way to produce fuels, or burnt directly as fuels.

4. Farm waste. Dairy, poultry, the meat-packing industry, and other farm operations also produce substantial quantities of waste. These wastes are varied in their nature. Poultry waste is mostly cellulosic in the form of solid chicken litter plus the chicken excrement. Dairy waste is often in the form of a watery slurry. Meat operations produce waste that is rich in fats. All of these wastes have useful energy content, and each requires a process that is suited for its utilization.

At present, most biomass is used directly to produce heat, which in turn could be used to make steam. Since biomass often has high moisture content, its combustion is not very efficient—one ends up producing a lot of steam on the wrong side of the heat exchanger. It is a common practice to get rid of most of the moisture first by air drying. The heating value of air-dried biomass tends to be around 7,000 Btu/lb, which is roughly half that of bituminous coals and only about a third that of crude oils. For this reason, direct combustion of biomass is not likely to replace pulverized coal combustion for power production in any significant amount. Slow pyrolysis, or torrefaction, is another process by which most of the moisture and smoke from biomass is removed to produce a carbon-rich char. This char, which is typically about 70% of the weight of the air-dried biomass, retains about 90% of heating value and can be fed along with coal into a combustor or a gasifier. The synthesis gas produced by the gasifier is a mixture of carbon monoxide and hydrogen. It can be either burnt in a gas turbine to produce power, or catalytically converted to produce other fuels such as methanol, gasoline, or diesel.

When the water content of biomass is very high, such as with animal waste from dairy farms, the preferred path for recovering energy is through anaerobic digestion. This bacterial process occurs in the dark and in the absence of air. The bacteria produce biogas (mostly methane), which can be purified to the pipeline quality of natural gas, or burned locally to provide heat or electricity. The process is most often used to manage manure, but recent advances make it possible to convert other agricultural and food waste streams by this method. It is also used in certain water treatment facilities, where the wastewater is first treated with aerobic bacteria or algae, and the resulting biomass is digested anaerobically. The methane gas produced by digestion is then used to produce electric power and to offset the power requirements of running the facility. The anaerobic digestion process also occurs in landfills, and in most instances that methane gas simply escapes into the atmosphere, contributing to the greenhouse gas pool. While important as an adjunct to other processes, the contribution of anaerobic biogas to our total energy needs is currently insignificant.

Global Biomass Potential

There are many sources of information and methods that can be applied to estimate the potential of global biomass to supply energy. We consider several approaches to this estimate below. The various approaches include considering how much biomass could be grown based on the amount of energy available from the sun, the amount of arable land, the types of crops available, and the prevailing climate conditions.

One approach to estimate the potential of biomass is to consider how much plant growth the sun's insolation allows. Land-grown, high-energy-producing crops such as corn grain and sugar cane require temperate to tropical climates, fertile soils, and water—factors that further limit areas suitable for their growth and energy production. The photosynthesis process is an inefficient converter of solar energy, and plants use much of the energy originally produced for their own survival when the sun is not available or temperatures are low. The theoretical maximum efficiency of the photosynthesis pathway in utilizing sunlight is estimated to be 10%, and the observed average efficiency of land-based plants is only 0.25%. If we use this number in conjunction with the 23,000 CMO of annual insolation over the globe and the fraction of Earth's surface that is land, we can estimate the annual biomass production on land as 14 CMO. This top-down calculation serves as an extreme upper bound to the amount of biomass that can be produced globally.

Another way of estimating biomass availability and its energy potential is to examine the production of cereal grains, which tend to contain most of the energy available from those crops. Their production will be limited to temperate and tropical climates with arable land. The total area of the temperate and tropical areas of the world is approximately 50 million mi^2, and about 10% of this is arable. Of this 5 million mi^2 of suitable land, about 1.7 million is devoted to the production of grains that today are used almost exclusively for food produced for direct human consumption, or as feed for animals that will be consumed by humans. The yield from these 1.7 million mi^2 is currently about 1.9 billion tons. If we further assume that grains generally have an energy content of 8,000 Btu/pound—roughly that of semidried wood—then the energy in the grains would correspond to 0.21 CMO—roughly equivalent to an average per capita food consumption of 3,000 Calories per day for each human on Earth.

If we extrapolate the same productivity to the remaining 3.3 million mi^2 of suitable land, we estimate an annual global production of grains of 5.6 billion tons, with an energy content of 0.58 CMO. These grain products are generally accompanied by larger amounts of waste material, much of which can be (or is) recycled to maintain soil quality. The rest might be usable as a source of energy. Even so, the total amount would be no more than 0.7 CMO, or about 1/20th of the estimate we arrived at by the insolation availability approach above.

A joint study by the U.S. Department of Energy and the U.S. Department of Agriculture considered the potential for using various biomass feedstocks, including not only starch, sugar, and oil crops, but also animal waste, agricultural waste, grasses, and forest residues. Based on some reasonable, albeit optimistic, assumptions, the report concluded that the United States could produce more than a billion dry tons of biomass annually from all of these resources in a sustainable manner. A dry ton of biomass typically has a heating value of 14 MBtu, so a billion tons would correspond to slightly less than 0.1 CMO, which is only about one-seventh of the U.S. primary energy consumption.

Useful amounts of biomass can be produced only in limited regions of the world. Not only is adequate sunlight essential, but adequate water, a conducive temperature regime, and proper conditioning of soil are equally important factors for useful plant growth. Arctic regions, many arid and mountainous areas, and oceans are not suitable for intensive biomass production. Keeping these limitations in mind, Christopher Field and his associates at the Carnegie Institution examined the natural plant productivity of various land resources such as crop lands, pastures, and abandoned crop lands and pastures.[18] They concluded that globally, the amount of biomass produced is equivalent to about eight times the amount of global energy consumption, and thus in principle offers the possibility for meeting global energy needs. However, if biomass energy is to be used with the objective of minimizing greenhouse gas emissions and the attendant climate change, then biomass must be grown and harvested in a manner that (a) does not conflict with food production, and (b) requires minimal energy input for irrigation and fertilizing. If we now restrict the use of biomass for energy to only the abandoned areas (so as not to compete with other land uses) and consider the natural plant productivity in these areas, the global yield would meet only about 5% of the current energy demand. We note that this analysis did not address biomass growing in water, and that oceans cover more than two-thirds the surface of Earth. Crops such as giant kelp and microalgae can be grown in oceans, but the problems of growing and farming them are greater, and consequently, the costs of the products are higher than those from land-based crops.

To summarize, estimates for global biomass potential for energy production on land range from 0.5 to 14 CMO. These numbers represent anywhere between a sizable fraction of the total global energy demand to several times that, meaning that biomass will likely play an important role in future energy production.

Factors Limiting the Use of Biomass

Transporting biomass from where it is cultivated to where it is needed is an important factor that limits the use of biomass for energy. The economic limit for transportation of biomass by truck in the United States is about 50 miles. Biomass can also be converted into gaseous fuels. The range for transporting gaseous products is perhaps 100 miles for a low-Btu product, and 350 miles for

18. Field, C. B.; Campbell, J. E.; Lobell, D. B. Biomass energy: The scale of the potential resource. *Trends Ecol. Evolut.* 2008, 23 (2), 65–72.

a medium-Btu product. The high-Btu product, often called substitute natural gas, can be distributed over distances equivalent to those practical for natural gas. Many biomass products (e.g., ethanol and methanol) are liquid fuels that are easily transported. However, producing liquid fuels from biomass incurs an energy loss of 20–30% in the conversion process. Other losses of energy can occur if the biomass is not used shortly after harvest. In some cases land contours prohibit economically productive farming or forestry. These factors relating to transportation create significant limitations on the availability and utility of biomass on a global scale.

Wood

Wood is most often used as a direct source of heat with widely varying efficiencies—from perhaps less than 10% for an open fire to 30–40% for tandoori ovens, and even higher for specially designed wood burners. But neither heat nor wood itself is easily transported.[19] To make the energy in wood transportable, it must be converted into electricity or liquid fuels. Production of electricity from wood using advanced gasification/combined-cycle technology will have an efficiency of about 40%. Conversion to methanol is about 57% efficient. Thus, for example, nearly 6 million mi^2 of Brazilian eucalyptus plantation could be expected to produce 1 CMO/yr of methanol fuel.[20] If a further step were to be taken to produce gasoline directly from wood, then the total land requirement would be more than 8 million mi^2 of plantation forest to produce 1 CMO/yr.

Because the production efficiencies associated with energy from biomass are small, the land requirements are large. This factor grows still further in importance when one considers the competing uses for the best land, including food and fiber production, living space, infrastructure, and habitat for an entire ecosystem. Unless productivity of all crops is increased and efficiency of food distribution is improved, the world will need to devote even more than its current use of arable land (about 5 million mi^2) to annual crops, to satisfy the needs of a growing population for food. In addition, a half-million square miles are now devoted to permanent crops, such as fruits, nuts, vines, and natural fibers (e.g., cotton). Growing prosperity in the emerging economies of the world is leading to an increased use of "higher value food" such as chicken, pork, and beef, all of which require growing even greater quantities of grains. Still other land is needed for nonfood uses, such as wood grown for pulp and

19. Conversion of the wood energy to a transportable form is generally needed because the quantities of this energy available at a single location are larger than most prospective onsite uses. Timber is usually transported only a few miles before it is used, although it can be rafted down rivers for substantial distances. It may also be economically feasible to move pelletized wood much further than raw wood, depending on the transportation services available, but the pelletizing process adds costs and requires energy.

20. Natural forests might need to be two or more times as big to produce 1 CMO/yr.

paper production, construction materials, and preservation of as natural forests. Given that the total arable land area is fixed, all needs call for increased productivity. Safeguards against nutrient replenishment and soil erosion will be integral parts of programs aimed to grow more food. Grown sustainably, biomass is a renewable resource whose use takes precedence for food, clothing, and shelter before it is used for energy.

Recovery of Biomass Energy from Wastes

In contrast to requirements for dedicated energy crops, the land requirements for production of energy from residues are minimal. It is important to estimate how much energy could actually be recovered from these "wastes," given that the most easily extracted energy has usually already been obtained for another purpose such as food. Current estimates of residue production suggest a recoverable energy content of about 0.4 CMO/yr, but little of this energy is actually recovered at present. Estimates of future usable biomass residues range from 0.17 to 2.8 CMO/yr. The lower limit corresponds to the lowest estimate from a "demand driven" scenario, and the upper limit is from a "research potential." Many waste sources may be exploited. In particular, bagasse (the spent material left after sugar juice is squeezed from the cane) is a relatively large potential income energy producer.

Current sugar production results in the rejection of an estimated 350 million tons of bagasse annually. Even this seemingly large amount has an energy content of only 0.02 CMO. Other sugar cane residues left in the field, which would require extra collection efforts and costs, could supply perhaps an additional 0.01 CMO/yr. The value of this extra 0.01 CMO is questionable, given that many of the residues need to be tilled back into the soil where they can perform the important role of soil reconditioning.

Much of the potentially available bagasse energy is now used to supply energy for processing the sugar—for heat to concentrate the sugar syrup, and for electricity to drive the mill machinery. Wood wastes in lumber mills and paper-mill waste (black liquor) are similarly exploited for their energy content. Overall, a substantial fraction of the waste is used internally—probably more than half in the case of sugar, lumber, and paper mills.[21]

Currently, the energy recovered from *all other* agricultural residues and municipal solid wastes may be between 0.001–0.002 CMO/yr. The potential economic recovery may be no more than 0.005–0.01 CMO/yr.[22] If this amount

21. Such use could be viewed as a gain in energy efficiency in overall industrial processes; alternatively, the U.S. Department of Energy and some others count it as a biomass contribution to energy supply.

22. Data on energy production from these sources are fragmented and reported in so many different ways, for example, not by energy content but as tons of material or as electric power capacity, that it is difficult to calculate the actual quantities of energy produced.

of production were to be increased 10-fold, contribution of waste biomass to future overall energy use would still be very small.

Recovery of energy, especially electricity, from municipal solid wastes has many problems associated with conversion, handling, and transportation. Planning for pollution control is complex, given the variety of feedstocks. Residues are often toxic and treatment to render them environmentally benign is necessary. The collected refuse is bulky and the associated transportation costs high, so the conversion plants are frequently located in or near major cities. Proximity to plants has concerned the local citizens. With citizen opposition, and environmental and technical difficulties, relatively few waste energy recovery facilities have been built and operated. The primary advantage of their operation is reduction in waste volumes and therefore lower use of often-valuable land for disposal. In addition, operation of these plants can lead to the recovery of valuable material—for example, aluminum, glass, lead, and gold. In developing countries, villages are using biomass as well as animal and human wastes to produce biogas.[23] This low-Btu gas is a mixture of CO_2 and methane, produced by anaerobic digestion of vegetative and human wastes, and is used to supplement kerosene and other heat-energy sources. Production rates are low, and the quantity of this source will generally be miniscule with respect to global energy needs.

In contrast to direct use of plant biomass, use of waste biomass for energy production is not anticipated to play any significant role in meeting future energy needs.

Biofuels and Their Production

As mentioned earlier, biomass is gaining increased attention. Of all the income sources, biomass alone has the potential to provide desperately needed liquid fuels to the transportation sector. The idea of growing one's fuel as opposed to importing it from other countries makes a lot of sense at some level. Several different approaches are being developed to achieve the goal of producing transportation fuels from agricultural resources. These include production of biodiesel from various oil-bearing seeds or algae as a drop-in replacement for petroleum-derived diesel, ethanol from sugar cane or corn, and lignocellulosic sources for use as gasoline extenders.

One component of comparing the different approaches is to look at productivity, defined here as how many gallons of fuel can be produced annually from an acre of land. Comparisons could also be made on the basis of energy or water required for growing them. The cost of fuel is another key point of comparison. In principle a comparison of costs would capture all the factors

---

23. Biogas is being used as an unmeasured energy source in agricultural areas of developing nations and, with subsidy, by some cities in the developed world. Notable energy recovery facilities have been located in Chicago, on Long Island, New York, and in Paris.

Exhibit 7.7.
Biofuel Productivity of Various Sources and Area Requirements

Biofuel	Annual Fuel Yield (gallons per acre)	Annual Productivity (GO/acre)	Area Needed to Produce 1 CMO (thousand mi²)
Biodiesel			
Palm oil	3,140	2,800	610
Jatropha	1,440	1300	1,330
Rapeseed	980	880	1,950
Algae	5,000	4,500	154
Ethanol			
Sugarcane	4,050	2,700	633
Sugar beet	3,270	2,200	785
Corn	2,100	1,400	1,230
Wheat	1,760	1,200	1,450
Barley	780	525	3,270
Switchgrass	4,000	2,614	980

and therefore be quite relevant; however, most approaches are still in the research and development stage and so a comparison based on current prices would be unrealistic. For now, we will limit ourselves to land requirements. In exhibit 7.7, we list the productivity of the different approaches, and also the area of land needed at that productivity to produce 1 CMO of fuel.

Except for algae,[24] most biofuels require hundreds of thousands to more than a million square miles for one CMO. To put that figure in perspective, the total land area of the United States is about 3 million mi². The agriculture-rich states of North Dakota, South Dakota, Nebraska, Kansas, Minnesota, Iowa, and Missouri together encompass an area of a little more than half a million square miles. More relevant than total area would be a comparison with the cropland that is currently used for growing grains and cereals. The answer would depend on the type of crop and the country where it is grown. If we use the number for corn in exhibit 7.7 (1.2 million mi²/CMO), then producing 0.025 CMO of ethanol to displace 10% of U.S. annual oil consumption would require 30,000 mi², or 6% of the U.S. cropland area. According to the Worldwatch Institute, the fraction of cropland area needed by various countries to displace 10% of their transportation fuels varies from 70% in Europe

24. Different sources give productivity numbers for microalgae that vary by as much as a factor of 20 in either direction of the value provided here. Many sources cite values based on data from growing algae for micronutrients; others estimate values from small-scale energy-related experiments. The algae technology is currently under development.

to about 3% in Brazil.[25] Their estimate for the United States is 30% of the current cropland. Their higher proportion of cropland probably takes into account regional variability in productivity and is probably more realistic.

Biodiesel

Diesel engines can use vegetable oils, and indeed, Dr. Rudolf Diesel ran his original engines on peanut oil. Seeds from palm, soy, rapeseed, coconut, jatropha, or almost any oil-bearing plant could be used for this purpose. According to a study comparing the productivity of various oil crops, palm produces the largest quantity by far of oil per acre of land. The productivity of palm is 635 gallons per acre, followed by coconut (287), jatropha (194), rapeseed (127), and soy (48). The high productivity of palm would make it the plant of choice for this purpose. The limiting aspect of palm is that it grows well only in the tropics. In the United States, biodiesel is produced mostly from soy oil. Of these oil crops, soy places the largest burden on land requirements. In Europe, most biodiesel is produced from rapeseed.

Chemically speaking, the vegetable oils expressed from the seeds of plants—often referred to as straight vegetable oils or SVO—are triglycerides. Triglycerides have three molecules of long-chain fatty acids (each containing 14–20 carbon atoms) attached to one molecule of glycerine. This structure makes SVO more viscous than petroleum-derived diesel fuel, and its direct use often requires a separate fuel tank and auxiliary heating system to warm the SVO so it becomes sufficiently fluid. SVO is more typically converted into biodiesel by treating it with methanol in the presence of a catalyst including enzymes. The net result is three smaller molecules of fatty acid methyl esters (FAME). Glycerol is a by-product of this reaction. As demand for biodiesel has increased, so also has a surplus of glycerol been created in the market. Many investigators are looking for ways to use this glycerol.

The properties of biodiesel are very similar to those of diesel fuel, and biodiesel can be blended with diesel in all proportions. The chief difference between the two arises from the presence of oxygen in the biodiesel, which reduces its net heating value from 19,000 Btu/lb to 16,000 Btu/lb. The presence of oxygen also has some beneficial impact on the reduction of tailpipe emissions of unburned hydrocarbons, carbon monoxide, and soot.

Of course, the main reason for using biodiesel is that as an income resource, its use could lead to reduced overall emissions of CO_2. The impact of biodiesel on greenhouse gas emissions is not as straightforward to assess as it might appear. It depends on *how* and *where* the fuel was grown, and we address that question later in this chapter after reviewing the major sources and practices pertaining to biodiesel and the other major biofuel, ethanol.

25. *Biofuels for Transport: Global Potential and Implications for Energy and Agriculture*, Worldwatch Institute, London, 2007.

The global market for biodiesel has grown vigorously, largely in response to tax incentives, carbon credits, and renewable fuel targets mandated by many governments. In 2005, biodiesel production in the United States amounted to more than 75 million gallons, up from 23 million gallons in 2004, and is slated to increase to more than a billion gallons by 2012. Production in the European Union has also grown steeply, exceeding 1.8 billion gallons in 2006. Demand in Europe has been very strong because the European Union mandated incorporation of 5.75% alternative fuels by 2010. Regulations in a free market economy create the situation of *mandated demand,* and European production alone cannot meet that mandated demand for alternative fuels at present. Thus, most of the biodiesel produced in the United States, parts of south Asia, and Africa is shipped to Europe. In view of these mandates, biodiesel plants are being installed all over the world. To serve the market for biodiesel, farmers are increasingly planting oil-bearing crops.

According to a recent market survey, *Biodiesel 2020,*[26] between 2004 and 2007 the global installed capacity for biodiesel production increased from 660 million gal/yr to almost 7 billion gal/yr. The installed capacity was projected to grow even higher very quickly, and the number of countries producing biodiesel to increase from 20 in 2007 to more than 200 in 2010. As would be expected, this steep increase in capacity is running into resource limitations. This is evidenced by the fact that while global installed *capacity* to produce biodiesel increased from 660 million gal/yr to almost 7 billion gal/yr, the actual global *production* over the same period increased from 600 million gal/yr to only 2.7 billion gal/yr. Countries such as India, Malaysia, Indonesia, Mozambique, and Brazil have plans for large commercial projects for the production and export of biodiesel, encompassing hundreds of square miles of land for cultivating jatropha.

To put this production in perspective, while the 2.7 billion gal/yr correspond to more than 10% of annual global vegetable oil production, they represent less than 1% of the annual global diesel fuel consumption. Even the projected 7 billion gal/yr would meet less than 3% of the current demand for diesel.

Ethanol

Ethanol is the other important biosourced transportation fuel. Ethanol's chief advantages are that it can be blended into gasoline in up to 10% by volume, and at that concentration it can be used in the current fleet of automobiles without any modifications. To use a higher proportion of ethanol than 10%, the engine would need to be modified. As a blend, ethanol boosts the octane number of gasoline and also reduces hydrocarbon emissions from the tailpipe.

26. W. Thurmond. *Biodiesel 2020: A Global Market Survey, Case Studies, and Forecasts,* 2nd ed., Emerging Markets Online, Houston, Tex., 2008.

These benefits combined with the desire to grow one's own fuel have led to a strong push for promoting ethanol as a fuel. The push came in the form of legislative actions for reformulated fuels standards, requiring oxygenates, in particular, ethanol. The push also manifested as subsidies provided to farmers and fuel blenders. Global production of ethanol saw a steep rise from about 1 billion barrels in 1996 to 14 billion gallons in 2006. There are several paths available for production of ethanol.

Sugar-Sourced Ethanol

Fermentation of sugars by yeast produces ethanol and is an ancient process. Many sugar-rich plants have been used to produce ethanol: sugar cane, sugar beet, and sweet sorghum are prominent. Sugarcane is possibly the best choice for energy yield. The main technical challenge in this fermentation process is that the yeast, *Saccharomyces cerevisiae,* cannot tolerate ethanol concentrations above about 12%. This limited ethanol concentration translates into a higher energy cost for distilling ethanol away from water. The heat for distilling ethanol from sugarcane is generally provided by burning the cane residue, called bagasse, after the juice has been expressed.

The greatest success with ethanol has been in Brazil with sugarcane. Following the oil embargo of the 1970s, Brazil embarked on an aggressive program of producing ethanol from sugarcane juice with the objective of reducing the country's dependence on imported oil. Brazil undertook several measures: it mandated 20–25% ethanol blends in all regular gasoline, promoted development of flexible-fuel vehicles that could use any blend of ethanol and gasoline, undertook policies to help farmers reduce the cost of sugarcane, and provided incentives for distillers to convert the cane sugar into ethanol. These measures were necessary to overcome the financial difficulties that the sugarcane industry faced in the 1990s, when the price of competition product—gasoline from crude oil—declined sharply and the sugarcane industry could not sell its ethanol. Through these measures, Brazil has achieved a position of global leadership in ethanol production. Brazil's ethanol industry is now mature enough that it can produce ethanol at half the price of gasoline on a volumetric basis. When this value is corrected for the differing energy contents, the price of ethanol in Brazil is still about three quarters that of gasoline.

Starch-Sourced Ethanol

Many plants, such as the grains corn, wheat, and barley, are rich in starch. Starch is a polymer of glucose and can be easily hydrolyzed into sugars, which can then be fermented into ethanol. Thus, starch plants are potential sources for ethanol, particularly for countries in the temperate zone where these plants can grow well but sugarcane cannot. Corn is the preferred source for ethanol in the United States. Corn has the highest starch content of the various crops grown in the United States, and about 15% of the corn grown there goes to producing ethanol. Its cultivation was also promoted to provide assistance to farmers in the midwestern states.

The process for producing ethanol from corn consists of first milling the grain, either in a wet or a dry process, to separate the cellulose, proteins, and starch components. The wet-milling process is more expensive but allows for coproduction of high-fructose corn syrup used in the processed food industry. The dry-milling process is less expensive but affords a less-valuable animal feed as the principal coproduct. After milling, the starch is treated with amylase enzymes to produce sugars, and finally these sugars are fermented by the yeast.

The appeal of using biofuels is that by this route we use the plants ability to capture the sun's energy in producing the fuel. However, the overall process for producing ethanol requires not only the sun but also energy inputs in the form of fertilizers, cultivation processes, hauling, milling, and distillation. A much-debated issue has been the amount of energy obtained from ethanol relative to the energy invested for its production. Depending on the assumptions made, various analysts have concluded that this ratio can range from less than one (about 0.7) to about 1.3. The energy return on energy invested—more specifically, the energy return on the *fossil energy* invested because solar inputs are not included here—swings from being negative to positive when one takes credit for the coproducts. The value of coproducts is an important consideration in industrial processes, including petroleum refining and any biofuel production.

Fossil energy inputs for producing ethanol are natural gas (for fertilizers) and coal (for distillation) and a relatively smaller amount of petroleum (for diesel fuel for tractors). Thus, producing ethanol has some merit in reducing oil consumption since it is a way of using energy from coal and natural gas in the transportation sector. The net return, however, is at best only 30%; for 100 units of fossil energy invested as coal and natural gas we can get at best 130 units of energy in ethanol. In comparison, the net energy return for ethanol production by the sugar cane process is 8 times or 800%. Therefore, in terms of ability to add to the total energy pool, ethanol production from starch is marginal at best.

Cellulosic Ethanol
Also referred to as second-generation biofuels, cellulosic ethanol uses nonedible resources of plants, including corn stover and the residues from sawmill operations and forest trimmings. Other sources could include annual and biannual energy crops such as switch grass, or rapid-growth trees that can be harvested several times at four- to five-year intervals, such as eucalyptus, various poplars, sycamores, and willows.[27] The principal components of these materials are lignin, cellulose, and hemicellulose. Like starch, cellulose is also a polymer of glucose and other six-carbon sugars, and hemicellulose

---

27. Perhaps four or five separate crops can be coppiced over a 20- to 25-year period before the land loses an important fraction of its original productivity and must be either abandoned or provided extensive renewal to return it to an economically viable production facility.

is a polymer with five-carbon sugars. Lignin is a complex polymer that wraps around these carbohydrate polymers, and the resulting composite imparts structural integrity to plants and trees. At the same time, the composite structure makes lignocellulose resistant to hydrolysis and other degradation processes necessary to extract fuel energy. For this reason, the process for producing cellulosic ethanol is more complicated than for sugar- or starch-based ethanol. It begins with a pulping process to separate the lignin and then converts cellulose and hemicellulose into sugars, which are then fermented to produce ethanol. However, the crystalline nature of cellulose makes it resistant to attack by most enzymes, and the amylase enzymes that readily convert starch into sugars are not suitable for cellulose and hemicellulose. Many of the advances in cellulosic ethanol have been in the area of biotechnology for producing more efficient cellulases (enzymes that hydrolyze cellulose) and strains of yeast that can ferment both five- and six-carbon sugars.

Will Ethanol Work?

Ethanol has been a priority of the U.S. government, which is paying midwestern farmers billions of dollars to grow corn for use in its production. The program is experiencing many difficulties, the process has yet to be perfected, and the government's original estimates that ethanol would be relatively cheap are proving incorrect. Around the globe, nations that have considered blending their gasoline with ethanol are wary due to the cost ineffectiveness of the American program. Also, greenhouse gases are released in this process from the use of fertilizers, farming, and the transport of corn to locations where it is transformed into ethanol. On the other hand, the Brazilian program is performing well, since growing and processing sugarcane to ethanol is not as energy intensive as the corn approach.

Butanol

Fermentation processes can also be used to produce butanol, which like ethanol is an alcohol, but butanol has four carbons instead of the two in ethanol. The extra carbons make it more compatible with gasoline and also increase its heating value to about 90% of gasoline (ethanol's heating value is about 67% of gasoline). The production process for butanol would use the same feedstocks as ethanol, but instead of yeast it would use *Chlostridium butylicum*. While largely untested to date, theoretically butanol can be a drop-in substitute for gasoline and therefore will not need development of new engines that could use it. Butanol is still in the developmental stage, meaning that it is years away from being cost-effective enough for widespread use. Research continues, because butanol has the potential to revolutionize the transportation industry given its cost-effective production from biomass.

Algae-Based Fuels

A promising potential source of energy is microalgae, which are extremely small seaweeds. All microalgae contain some oil that can be extracted and converted into fuel. Scientists are currently breeding strains of algae with very high oil content in order to extract as much oil as possible. Algae can be used to produce different kinds of fuel, including butanol, ethanol, and biodiesel. Biodiesel is the fuel most commonly produced, due to the widespread use of diesel engines.

The chief advantage of algae is that it has an exceptionally high yield of oil per area, up to 5,000 gal/yr per acre. This yield is many times greater than for any other biofuel. At that productivity, it would take 154,000 mi^2 to produce 1 CMO a year (compared to about 1.2 million mi^2 of corn). Algae are not difficult to grow and can thrive in saltwater or even our wastewater. When grown in wastewaters, algae provide an additional benefit of removing nutrients for growth from the waste streams. The benefit is that these nutrients from agriculture runoff can have deleterious effects on the environment by leading to the formation of "dead zones" in the seas. Algal farming does not require premium agricultural land and can use marginal lands, or even the seas. These benefits in energy produced and land usage are accelerating the pace of algae research. Algae are plants and therefore require proper nutrients and protection from dangers, such as bacteria. Researchers have found that in addition to focusing on developing strains of algae to produce the most oil, they must address the need for these potentially weaker strains to grow properly and stay healthy.

Algae-produced fuels cost $15–30/gal and are not yet economically viable. They are rapidly becoming less expensive to produce, so it is likely that algae could become a significant source of transportation fuel. Research is needed to drive down the cost at just about every stage of the process—growing, harvesting, processing—as well developing useful co-products. Many companies have announced plans to develop algae for fuels using innovative approaches. Some deal with new strains that can harvest sunlight more efficiently, and others deal with new ways growing and harvesting them. The Wall Street Journal recently reported plans of a small start-up, LiveFuels, to grow algae offshore, grow fish on them, and then harvest fish from which they will produce the oil.[28]

The Environmental Impact of Biofuels

Biofuels have been developed with an eye to ending our dependence on fossil fuels and helping the environment. Granted that if they are successfully and

---

28. Russel Gold, "Biofuel bet aims to harvest fish that feed on algae," The Wall Street Journal, August 18, 2009.

economically developed, the first provision could come true. But what about the second? Will they really help the environment as much as claimed? The answer is a resounding "It depends." While biofuels emit less greenhouse gases when used in cars, the steps taken to produce, refine, and transport them do produce greenhouse gases. Careful analysis has shown that the energy investments in terms of fertilizers and thermal processing can be substantial, and if these inputs were coal-derived, the *total* greenhouse gas emissions would be larger than when using petroleum as fuel.

The need to change our land usage to support the production of biofuels also causes emission of large quantities of greenhouse gases, during the first decade. If a rainforest were cut down to make room for planting palm trees to produce biodiesel, it could take more than a hundred years before the cumulative savings, from switching to biodiesel from petroleum diesel, would equal the immediate release of carbon to the atmosphere caused by removing the forest.

Conclusion

Let us again consider the potential of each source to provide energy. Geothermal energy has significant potential, due to the large amounts of suitable hot dry rocks. Geothermal energy can potentially provide up to 2 CMO/yr. In contrast, hydropower has a very limited potential for growth because many of the suitable sites have already been dammed. Due to the uprooting of people and destruction of regional areas caused by dams, there is much opposition to further expansion of hydropower, and its potential for growth is limited to, at most, doubling from the current level, to 0.4 CMO/yr.

Wind and direct solar have very high potential for meeting future energy needs. The discrepancy between what is potentially available (23,000 CMO annually from the sun) and what we use today (about 0.2 CMO) is truly staggering. For example, today, we commercially use only 1% of the 20 CMO of global biomass. In turn, even 20 CMO of biomass represents a miniscule fraction of the direct solar energy striking Earth. Wind and concentrated solar thermal are both priced competitively with fossil sources and have extremely high potential if harvested fully (>10 CMO/yr). Yet wind and direct solar currently contribute less than 0.005 CMO annually to global energy. Of the various income energy sources, they have the best capture rates! The obvious questions are, *Why not more? Why not now?*

The answers to both questions lie in logistics: the inertia embedded in large systems and the variability of the resources—variability that often does not match well with human needs. Other factors include their diffuse nature and remote locations. Wind and direct solar will require tens of thousands of square miles to produce CMO-scale energy.

Development of wind at CMO scales would alter the landscape radically. Areas good for wind power are often in the path of migratory birds, including many endangered species. The society must collectively decide if we are willing to accept the extinction of some of the species. Increasing wind power, which is intermittent, to meet more than 20% of total electricity demand would require building a comparable amount of backup facilities—possibly natural-gas-fired plants—to ensure a constant energy supply. These challenges will make it difficult to realize the full capacity of wind power, but even with the difficulties we can still expect this source to grow to provide energy at a scale approaching 1 CMO/yr.

As for wind power, large-scale expansion of solar power to CMO scales would require integration into the electric grid. Infrastructural costs would be high, especially with photovoltaic technology. While it does not seem prudent to cover land with solar cells, taking steps such as building homes that (a) use the sun's energy effectively to provide light and heat, and (b) use active solar technology to provide much of their power are wise and useful ideas. Producing power from CSP systems is less expensive and lends more readily to integration at utility scale. Its expansion, like that of wind, will be determined largely by the choices society makes with respect to development in some environmentally sensitive areas. The sun provides an immense amount of energy, and our goal must be to ultimately harvest as much of our needs from it as we can.

Finally, there is biomass and the biofuels that can be produced from it. As we have seen, they are hardly perfect, whether economically or environmentally. Future developments could add to the feasibility of renewable energy utilization. For example, developments in the fields of plant biology and biochemistry now being pursued could result in increased plant yields and more efficient biomass-to-energy conversions. Investigations related to improvement of biomass yields, and even the utilization of biological processes to produce useful electricity, are in progress.[29] Even so, these improvements might only reduce area requirements by a factor of 2 or more. Newer developments of genetically modified foods—whose use is opposed by some—have provided crops with resistance to disease and pests, and even the ability to resist herbicides, thereby increasing crop yields. Also, ways might be developed to effectively use nonagricultural lands, such as marshes, to grow reeds for energy. To date, such attempts have not been very successful.[30]

Storage of energy from income energy sources is also a major requirement, with trade-offs. Hydropower can store energy until it is needed. Biomass can in principle also store energy, but both biomass and hydropower have limitations related mainly to area requirements. Solar thermal power systems are not nearly as demanding of area, and solar thermal systems can store energy as heat for a few hours to mitigate the daily variations in availability and demand.

29. Worldwide improvements in agricultural practices, as well as genetically improved grains (e.g., rice), have about doubled the global yield of grain per unit of land farmed over the past four decades.

30. Use of aquatic plants (e.g., water hyacinths and giant kelp) has been investigated, but these efforts have not resulted in attempts at large-scale development.

Renewable energy sources offer an opportunity for the world to begin transitioning away from its dependence on fossil fuels. There are factors limiting the growth of each of these sources, but when combined, they have the *potential* to transform the composition of the world's energy production in 50 years. As a result, it is wise for nations of the world to invest in renewable resources, which are our income sources, depending upon their specific circumstances and the suitability of their land for each potential use. Use of income sources will depend on several factors:

- Relative cost of the energy produced by the various income sources compared to that of the inherited sources
- Future technical and engineering developments
- World interest in development and deployment of income sources, as expressed in laws, regulations, and *initial* subsidies that favor their use
- Overall patterns of energy demand, which could range from 4.3 to 9.4 CMO/yr in 2050 (as projected in chapter 4)[31]

The business-as-usual projection requires that we triple our energy production by 2050, and the projection with assiduous conservation efforts requires that we double our energy production by 2050. It is apparent from the status of renewables availability, research, and global policy considerations described in this chapter that placing emphasis only on energy income technologies will be insufficient to avoid early loss of our inherited resource base. Improving the efficiency of providing what we need combined with learning to use less energy and associated goods and services, will be necessary if we are to avoid catastrophe.

31. We do not calculate overall income energy contributions for the lowest 2050 projection, 2.0 CMO/yr, because that scenario is considered the least likely and also most complex to estimate.

Part III

The Path Forward

8

Energy Efficiency and Conservation

Previous chapters in this book focus on the production of energy from different sources and how we might increase the supply to meet the anticipated growth in demand. In this chapter we focus on options to manage the energy demand. There are many ways—other than complete avoidance of the use of goods or services that demand energy—by which we can "save" energy; actually, we are not saving but reducing the growth in the demand of energy. It is often convenient to think of savings arising from two categories: energy efficiency and energy conservation. Energy *efficiency* reduces the energy necessary to perform a desired task, and energy *conservation* includes all actions that avoid unnecessary use of energy. To use the automobile as an example, development of techniques that reduce the fuel needed to go from one place to another is an example of improved energy efficiency. Substituting the automobile with a more efficient mode of transportation or the avoidance of the activity entirely would be examples of energy conservation.

Thoughtful use of both conservation and efficiency will be necessary if we are to achieve substantial reductions in our future energy use as individuals, nations, or the world as a whole. As discussed in chapter 4, the global energy use projected for 2050 under three scenarios with three differing growth rates ranges from a high of 9.4 CMO/yr to a low of 3.9 CMO/yr. Our recent energy use of approximately 3 CMO/yr (since 2000) is on a growth curve that follows the trajectory of the high-consumption scenario. Improvements in energy efficiency have of course been made steadily over the past century and will likely continue in the future. Much of that improvement has already been taken into account in arriving at the projections for future growth. The 2.6%

annual growth in energy consumption has taken place notwithstanding steady improvements in efficiency. To bring the projected 2050 consumption down from more than 9 CMO, we will need savings that would not happen without a rededicated effort. That is why we say it will take a heroic effort through conservation and improved efficiency to keep that demand from growing to more than 6 CMO/yr by 2050. In other words, we would be counting on demand-side management to deliver the equivalent of about 3 CMO/yr in savings by 2050. By no means is that a small feat, yet we find that we will also have to expand the energy supply side by about 3 CMO/yr over our current rate of consumption to meet the demand of 6 CMO/yr. To the extent that much of our current 1 CMO from conventional oil is likely to be gone, the production side must develop additional energy sources capable of delivering about 4 CMO per year.

Large as the differences in annual consumption are under the three scenarios, the differences in cumulative consumption over the 50-year period are even more daunting. Total energy requirements of 270, 214, or 163 CMO for the three scenarios between 2000 and 2050 are truly substantial when we consider the potential limits to our inherited resource base. Recall from chapter 5 that global reserves *plus* speculative resources of conventional oil and gas are only 140 and 108 CMO, respectively. Coal reserves and resources at more than 1,000 CMO add up to numbers substantially larger than the projected cumulative energy use, as do some unconventional oil and gas sources. Income resources are potentially inexhaustible, but even substantial growth of income resources might not fill the gap in energy between our forecasted needs and our ability to supply that energy demand in 2050. In reality market forces will tend to counter a long-term shortage to occur. A shortage raises energy prices, thus encouraging both the expansion of our resource bases and suppression of the overall demand for energy—otherwise known as demand destruction. However, demand destruction can also occur through civil unrest and war or through natural calamities such as famine and epidemics. If we are to avoid those serious shortages that have the potential for developing into international conflicts over resources—something that the world has unfortunately witnessed many times with untold suffering—we have to start implementing measures for greater energy efficiency and conservation sooner rather than later.

Overview

In chapters 3 and 4, we reviewed how the world has been growing for many centuries, both in population and in demand for energy. But the rising global population explains only part of the growth in demand for energy. Most of the increase in energy use can be attributed to increased affluence, which has benefited those nations that are now highly industrialized, since the 17th century.

As affluence spreads to the rest of the world's inhabitants, we will most likely see increased use of global energy well beyond population growth. The growth in energy demand is most apparent in the two largest nations, China and India, whose populations now comprise about one-third of the world's total. Together their population is eight times that of the United States, but their combined energy consumption today is less than that of the United States. Many other countries with relatively large populations are on the cusp of experiencing broad prosperity (assuming, of course, the availability of energy to support that prosperity increase), and therefore increase in demand for energy supplies will, quite likely, be with the world for some time to come.

Many nations are attempting to expand their use of income resources that are in principle both large and inexhaustible.[1] Their attempts may be insufficient to provide the energy necessary to meet demand posed by the expanding world economy, unless substantial action is taken on a worldwide basis to curb our current and prospective future demands for energy. Yet, as we discussed in chapter 4, much of the increase in energy demand is coming primarily from nations that currently use relatively little energy. On what moral authority could developing nations be asked to forgo expanding their use of energy? Many countries are striving to improve the standard of living for their populations, including in some cases rising above near destitution to some modicum of a healthy decent living. We have to support that effort and bring billions of additional citizens of the world to fully participate in the global economy. Instead of only trying to curb their demand for new energy supply, we should be focused on helping all people to gain technology that makes the most efficient use of resources.

Since even significant voluntary energy savings from the developing countries will be inadequate (see below), the alternative for the United States and other nations with large per capita energy use is to (1) learn from nations who are more effective in their energy management, then (2) add to these efforts by developing and expanding newer and better conservation measures, and finally (3) encourage adoption of these efficiency and conservation measures in the world economy as a whole.

In the following pages we concentrate on past and future use of energy in the United States, because the United States is among the most profligate in use of energy and thus offers relatively higher potential for energy savings through improved efficiency and conservation. We begin our discussion with an overall historic look at global gains in energy efficiency. We then examine how efficiency of various operations might be further improved in the three main sectors of energy use: industrial, residential/commercial, and

1. Income resources are inexhaustible only if they are developed in a sustainable manner that allows nature to recover fully. Intensive biofuels production practices can sometimes deplete the topsoil at unsustainable rates, and therefore use of this income resource would not be inexhaustible.

transportation. We look at the various opportunities present in each sector, how large they are in terms of potential overall savings, and what research and development activities are likely to affect them. There is considerable interest these days in developing "green" technologies, and it is refreshing to see so many new and established companies making investments in this area. Many of them are likely to be successful businesses, providing handsome returns to their investors. However, if green technology businesses are to succeed as important global players, they will have to produce their products at a price that the developing countries can afford to pay. Consider individuals in the developing world who have improved their standard of living by purchasing sewing or weaving equipment, paid off by increased productivity. Can a similar model apply for say photovoltaic panels or wind turbines? Since much of the global increase in energy demand will be taking place in the developing countries, the green products must penetrate that market to have global impact.

Historic Improvement in Energy Efficiency

In physics and engineering, energy efficiency is defined as the amount of useful work derived from the consumption of a unit of energy. The concept has been extended to economics with the change that GDP has come to stand for useful work, and energy efficiency is measured as the monetary output (e.g., dollars) per unit of energy consumed. A related term, energy intensity, is the inverse of efficiency; it is expressed as energy units required per dollar of economic output. In physics, energy intensity often refers to the flux of energy per unit area. In this chapter, we use the terms "energy efficiency" and "energy intensity" as defined in economics and policy discussions. In so doing, we also note that this definition does introduce an element of distortion, because the value added by different activities—and hence the contribution to GDP—often has only an indirect connection with energy consumption. It takes far more energy to produce a pound of brass than to turn it into a faucet, yet the latter adds significantly more value and therefore makes a greater contribution to the GDP. Activities in the service industry, for example, banking or management consulting, are far less energy intensive than production of goods, yet they do contribute to the GDP.

The Earth Policy Institute has published historic gross domestic product for various nations and also the world as a whole (gross world product, or GWP). They use purchasing power parity as a basis for relating the productivity of different countries, which reflects the local living conditions better than straightforward currency exchange rates (i.e., market rate equivalency) would suggest. Because GWP will also increase as a result of inflation, the analysts express the GWP in constant dollars; in this case, they are related to the purchasing power of the dollar in 2005. We divided these inflation-corrected GWP data by the world's total energy consumption, which was taken from BP's

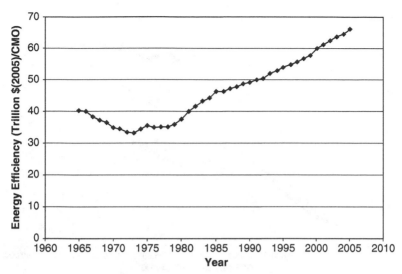

Exhibit 8.1. Global Energy Efficiency Has Increased During the Last Thirty Years

Statistical Review, and show in exhibit 8.1 how energy efficiency has changed during the 40-year period between 1965 and 2005 (in trillions of 2005 dollars of goods and services per CMO of energy consumed).

Earlier in this period, energy efficiency was declining and hit a low of about $33 trillion/CMO around 1975. After that it turned around and has been increasing ever since, to almost double its value by 2005. The turnaround came in the wake of the oil crisis and high price of energy in general. Increasing energy efficiency is of course good because it means greater productivity for the same amount of energy consumed. The really good news is that steady increases in energy efficiency continued even through the 1990s, when the inflation-adjusted energy prices were at their historic low point. This improvement occurred despite the fact that in some sectors, notably transportation, the efficiency gains of the 1980s were partially lost in the 1990s. In the 1990s, the strongest gains in efficiency use were observed in the industrial sector.

The global energy consumption in 1996 was 2.5 CMO and grew to 3.0 CMO in 2006, corresponding to a growth rate of 2.1%/yr. This growth in energy consumption occurred despite an increase in energy efficiency of 22% over the 10 years. Without those efficiency gains, the energy consumption for the same GWP output would have been nearly 3.7 CMO in 2006! Does this mean that efficiency gains saved us about 0.7 CMO in 2006? That would be true only if the world managed to have the same product output. It is unlikely that we would have such a high productivity without these efficiency gains. We note, however, that while greater efficiency enables increased output of energy (increased productivity), increased output of energy also spurs increased consumption of energy. Thus we caution against expecting commensurate energy savings from gains in efficiency alone.

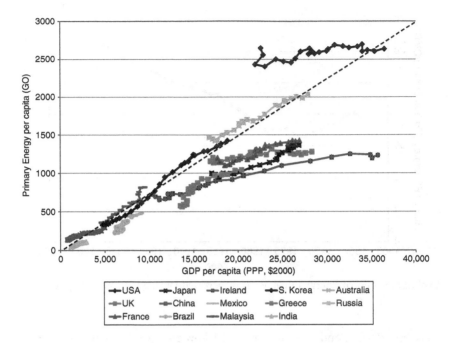

Exhibit 8.2. History of Per Capita Energy Use and GDP in Selected Countries

As shown by data in chapter 3, it is generally true that the per capita energy consumption of a country increases with its per capita GDP particularly for underdeveloped countries. We have also implicitly equated increases in GDP with the well-being of the population, and again, that statement is also by and large true. It is instructive to examine how per capita GDP and per capita energy use in different countries have changed over an extended period. Exhibit 8.2 does this for 13 countries from 1980 to 2004; the data are corrected for inflation and reported in 2000 dollars. Intercountry comparison was made on the basis of purchasing-power parity. The original data for this graph were culled from various sources by Paul Deninger of Jeffries and Company, a venture capital company in Waltham, Massachusetts.[2] We have modified the data to reflect our preferred unit for per capita energy consumption, gallons of oil equivalent (GO).

2. Private communication. Data presented at the Stanford Technology Ventures Program meeting in July 2008. http://alwayson.goingon.com/page/display/28097?param=session/326. Accessed August 2008.

The dashed diagonal line in the graph corresponds to an energy intensity of 10,000 Btu/$. Points above the line mean greater energy intensity, and those below the line mean lower energy intensity.

Viewing the points in the graph collectively supports the notion that increasing GDP correlates strongly with per capita energy consumption. However, this correlation is not a destiny and does not always apply. If we trace the historic evolution of GDP and energy consumption for various countries, we see several different profiles. Data for China, Brazil, Malaysia, South Korea, and Australia fall along the diagonal: they display a steady growth in GDP along with a concomitant growth in energy use; there is no change in the energy intensity of approximately 10,000Btu/$, and since a gallon of oil is equivalent to 130,000 Btu, the energy intensity of the dashed line corresponds to 1 GO/$13. Expressed as energy efficiency, it means that in all these countries it has taken 1 GO of energy to create $13 of value.

Next, we have countries like Ireland, Japan, and France showing a marked increase in per capita GDP but with a much lower increase in per capita energy use. The energy efficiency in these countries was close to $13/GO in 1980, but the increase in GDP since then has been achieved with much greater efficiencies. The United Kingdom and the United States show very little increase in per capita energy use during this 25 year period, yet their per capita GDP has grown substantially. Many analysts have used this breakdown of the correlation to suggest that efficiency measures can substantially reduce global energy consumption. We caution against this optimism.

The breakdown in correlation between energy-use and GDP reflects the nature of the economies of the respective countries. Some countries are clearly in the growing phase and relying heavily on industrialization. Others, like the United States, have been essentially in a deindustrialization mode characterized by a shift from a manufacturing to a services economy. Ireland's remarkable economic transformation during the past 20 years was enabled by its efforts to develop infrastructure and education facilities, which enabled growth of its semiconductor industry. This industry has much lower energy intensity than, say, steel or concrete manufacturing. At any given time the world as a whole cannot count solely on the low energy intensity of the services industry; someone must produce the steel, concrete, aluminum, glass, and other energy-intensive goods and thus supply the energy required for those operations. Thus, the need for ever increasing energy-efficient technology remains.

Recent Trends in the U.S. Energy Consumption by Sector

In chapter 3 we discuss the consumption of primary energy in different sectors. About 36% of primary energy goes into producing electricity, and the remaining 64% is divided between the industrial, residential/commercial, and transportation sectors. Exhibit 8.3 displays the total energy consumption in

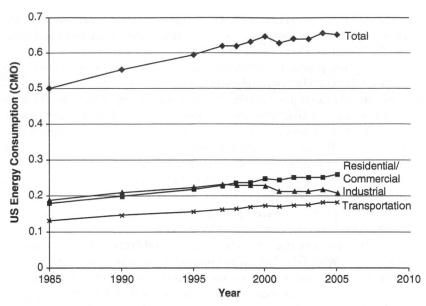

Exhibit 8.3. U.S. Primary Energy Consumption, Total and by Sector, 1985–2005

the United States between 1985 and 2005 along with its breakdown in the three major energy-use sectors of the economy. The consumption by sector in 2005 was as follows: commercial/residential sector, 0.26 CMO (40%); industrial, 0.20 CMO (32%); and transportation, 0.18 CMO (28%). According to data obtained from the U.S. Bureau of Statistics, the total energy use in the United States rose from 0.50 CMO in 1965 to 0.65 CMO in 2005. This general increase in energy use indicates that the lower energy consumption projected by the U.S. Energy Information Agency, World Energy Council, or by our own "fantasy" scenario (Case 4) that calls for the world overall to use less total energy beginning in 2030, cannot be achieved by increases in efficiency alone. It will also require widespread efforts in energy conservation.

Examination of exhibit 8.3 reveals several important trends. We can see that between 1985 and 2000 total energy use in the United States increased by a factor of 1.3, and has been almost flat since then. For reference, between 1985 and 2005 the U.S. population grew by a factor of 1.25. Thus, overall energy use in the United States has increased only slightly more rapidly than its population, and its per capita energy consumption has stayed flat over this period.

Energy use in the transportation and commercial/residential sectors has shown steady increases over the 1985 to 2005 period. Energy use in both sectors grew by a factor of about 1.4, with annual growth rates of 1.7% and 1.9%, respectively. Industrial energy use grew 1.2-fold between 1985 and 1997 and has been slowly declining since then. Energy use in the industrial sector is strongly influenced by competitive factors at home and abroad, and those pressures along with the general shift to the nonmanufacturing industries in the United States, have led to reduced energy consumption in this sector. Growth rates in the transportation sector were influenced by government regulation, notably the mandated increased fuel efficiency, as well as by market forces that reflected both this nation's general affluence and its infatuation with the automobile. The residential/commercial sector's energy use was mainly influenced by governmental promotions and incentives that resulted in use of more efficient household appliances and by regulations and incentive measures that resulted in greater energy efficiency in all new buildings and retrofits of some existing ones. We next take a deeper look at each of these sectors as we seek to identify and quantify opportunities for saving energy.

Industrial Uses

Over the past 45 years, U.S. energy use has grown at a rate almost equal to that of the country's population growth—1.03%/yr growth in energy compared to 1.15%/yr growth in population. In the United States, the industrial sector was the largest energy consumer until 1997. This sector is dominated by large companies that have access to large amounts of capital that could be deployed to affect changes in company and sector operations. The industrial sector is most likely to respond quickly to competitive forces and energy prices, as it did in response to the swings in energy prices accompanying the oil shocks of 1973–1974 and 1978–1979 when energy prices rose rapidly. Investments for improved efficiency compete with other investments within a company, such as product development, but because efficiency measures are often quickly paid for by the resultant cost savings, the companies are usually willing to invest in them. The initial responses are often mostly repairs to reduce unnecessary losses created by broken windows, leaky water and steam lines and valves, and a general attention to energy waste. In addition, the major energy consumers began the introduction of more fundamental approaches of energy management (discussed below). Some fuel switching also occurred that lowered costs.

Exhibit 8.4 shows the trend in value added per unit of energy consumed in U.S. industry between 1985 and 2004. It also shows the data for the manufacturing and nonmanufacturing (service) subsectors. The manufacturing industry is more energy intensive than the nonmanufacturing industry. In 1985 it added about $8 of value per GO while its nonmanufacturing counterpart added $18/GO. The flatness of the lines between 1985 and 1995 reflects the fact that relatively stable energy prices over this period provided little incentive for U.S. manufacturing to seek further reductions in energy use. Nevertheless, some improvements were made and the energy efficiency of U.S. industry rose, as measured by the product value per unit energy. The dramatic rise in the productivity of the nonmanufacturing sector after 1997 is noteworthy. Between 1985 and 1997, the nonmanufacturing subsector added around $30/GO, but since then has climbed to around $60/GO. It reflects the rise of first the Internet-based

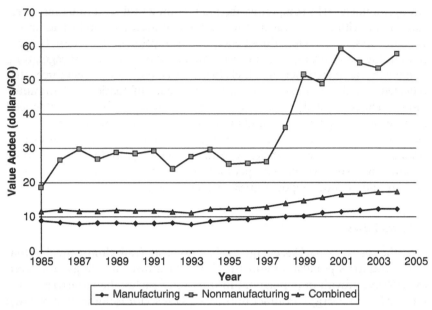

Exhibit 8.4. Value Added per Unit Energy Consumed in the U.S. Industrial Sector, 1985–2005

companies and then the financial sector. Yet, even in the manufacturing sector alone, the productivity improved by almost 50% between 1995 and 2005.

Recent studies by Ernst Worrell and coworkers at the Lawrence Berkeley National Laboratory conducted under the joint aegis of the U.S. Environmental Protection Agency's (EPA) Energy Star program and the U.S. Department of Energy, examined the potential for energy savings in two of the major energy consuming industries,[3,4] petroleum and petrochemical. The Department of Energy also sponsored other studies in the same vein to survey the industrial sector and develop a roadmap for reducing energy losses.[5] Even though U.S.

3. Ernst Worrell and Christina Galitsky. *Energy Efficiency Improvement and Cost Saving Opportunities for Petroleum Refineries: An Energy Star Guide for Energy and Plant Managers,* LBNL-56183, Environmental Energy Technologies Division, Lawrence Berkeley National Laboratory, February 2005.

4. Maarten Neelis, Ernst Worrell, and Eric Masanet. *Energy Efficiency Improvement and Cost Saving Opportunities for the Petrochemical Industry: An Energy Star Guide for Energy and Plant Managers,* LBNL-964E, Environmental Energy Technologies Division, Lawrence Berkeley National Laboratory, June 2008.

5. Energetics, Inc. *Energy Loss Reduction and Recovery in Industrial Energy Systems, Technology Roadmap,* November 2004, U.S. Department of Energy, Office of Energy Efficiency and Renewable Energy. http://www1.eere.energy.gov/industry/newsandevents/news_detail.html?news_id=8762. Accessed [January 2010.]

industry had already become quite efficient in the late 1990s, the studies identified many further opportunities that could together save more than 5 quadrillion Btu (0.034 CMO) of energy at prevailing energy prices. Surprisingly, these investigations found that the housekeeping deficiencies that account for a considerable amount of energy loss in some industries—leaky steam lines and valves, and so on—that were observed and corrected earlier, had now reappeared. The findings of the petroleum and petrochemical investigations also specifically identified approaches that could be applied to a wide variety of industries. Among the various suggestions for the petroleum industry was the conversion of more of the plant electrical system to a cogeneration mode, so that more of the plant's waste heat would be used to replace electricity previously purchased from a utility instead of being rejected to the atmosphere.

Another finding, applicable to several industries, was that companies could realize significant energy savings through use of heat pumps. Using low-grade heat that is often otherwise wasted, heat pumps can heat process streams more efficiently than combustion devices do. Heat pumps consume a fraction of the energy consumed by combustion heaters.

A major development in energy management, particularly in the chemical and petroleum industries, was the rearrangement of process piping so that necessary heat is first introduced to the process step requiring the highest temperature, and the degraded heat is then directed in sequence to process steps requiring successively lower temperatures ("pinching"). Once the heating fluid falls below a useful temperature, its temperature is raised by the injection of heat with heat pumps. This approach first became popular in the early 1970s, when high oil prices were first having an impact on the process industries. Also, the proper sizing of pipes and pumps to efficiently meet the flow requirements saved substantial amounts of energy. Likewise, energy savings were realized by proper, not excessive, illumination from high-efficiency lamps. Throughout these several studies, it was made clear that management responsibilities included surveillance of all manner of activities. These activities included review of equipment for proper function, for example, to assure that pulley drives were not worn or slipping, that valves were not leaking, or that enclosed spaces were properly sealed and not exposed to inclement weather. The studies concluded that each item may itself cause little loss in energy efficiency, but in aggregate they can lead to energy losses by several percentage points, causing loss of plant efficiency and profitability.

The study of the overall chemical industry showed that the largest losses are in off-site electricity production (at the utility) and its transmission to the plant site. The in-plant energy conversions, that is, the chemical processing steps themselves, have an industrywide overall estimated energy efficiency of about 77%. One reason for concluding that off-site electricity production offers the greatest opportunity for increased energy savings is simply that many individual energy-consuming components of the onsite processes have already approached their inherent limits of efficiency. Notable among these are the electric motors used to drive grinders, mixers, and pumps. The new ones have efficiencies of about 95%, leaving little room for further gains. On the other hand, the grinding, mixing, and pumping processes driven by these motors are themselves inherently inefficient. Even though these efficiencies are low, ranging from 5% for materials processing and refrigeration to 40% for pumps and fans to 80% for air compressors, ways to substantially increase their efficiency are difficult to envision.

A road mapping study by Energetics, Inc. included a discussion of the top 20 opportunities, tabulating both energy savings and the corresponding required investments. In many instances, the required investments were quite modest. The top 20 opportunities were grouped under five main categories, and the estimated potential savings under each (for the U.S. industries) are as follows:

1. Waste heat recovery from gases and liquids in the chemicals, petroleum, and forest products industries represent a major opportunity. Heat from process streams that are above about 300°F (150°C) is generally recovered using heat exchangers. Even so, streams with temperatures above 170°F (~75°C) carry about 0.05 CMO of energy for all U.S. industrial activities. Drying, calcining, and metal quenching operations provide additional sources for heat recovery. According to this report, about 0.012 CMO annually could potentially be saved through innovative recovery cycles.

2. Energy system integration involves a diverse set of methods for matching the heat sources and sinks at a plant. Together with implementation of best practices for steam and pump operations, this category can save the industry an estimated 0.009 CMO annually.

3. Improved boilers and fired heaters in chemicals and petroleum industries are identified as the next major category. Industrial boilers use about 6% of the total energy in the United States. Their efficiencies range from 60% to 80%. Super boiler technology combines innovations in high-efficiency burners, smart controls, and effective preheating among others to deliver efficiencies of 90% or higher with low emission of pollutants such as nitrogen oxides. Advances in heat exchanger technologies are also minimizing heat losses during various processes. Adoption of such systems could represent potential annual savings of 0.006 CMO.

4. Use of combined heat and power systems (cogeneration plants) instead of importing electricity is another major opportunity. Many industries require heat for their operations, and many power generation systems produce a large amount of waste heat. Recall that most gas- and coal-fired plants have a thermal efficiency of less than 40%. By colocating the power generation facility within the plant, the industry could effectively make use of the waste heat and save an estimated 0.005 CMO.

5. The use of sensors, controls, automation, and robotics can help maintain the plant to consistently meet product specifications. By reducing the

amount of off-spec product, this equipment can minimize energy use and cost. The report estimated a potential savings of 0.001 CMO annually.

This listing illustrates that the industrial sector by and large runs quite an efficient operation, and the opportunities that do exist are individually relatively small. The reports give the energy savings as 5,162 trillion Btu, about a quarter of the total 0.2 CMO used in that sector; in other words, the total energy savings attainable might amount to 0.05 CMO. So when looking to save energy approaching a CMO in the United States, we are not likely to find it in efficiency gains within the industrial sector.

Does this mean that we should not continue installing these improvements? Of course not! There is never any reason to waste energy, and many of the improvements also provide cost savings to the industry. These measures will likely be undertaken if the industry is convinced that fuel prices are likely to remain high for however many years needed to allow for the recovery of the necessary capital investments.

Change in the cost of petroleum fuel is the prime factor in the conservation-related attitude of U.S. industry. This is because petroleum-derived fuels are both a trendsetter for other fuel prices and the largest feedstock for the petroleum industry. A price signal through a carbon tax or a similar vehicle could spur funding to develop numerous technologies to make various chemical and industrial processes more efficient, but at this point most technologies developed for providing still further efficiency need to become cheaper before their use becomes widespread. As we just discussed, energy savings of 25% of the total U.S. industrial usage would be 0.05 CMO per year.

Residential/Commercial Energy

The residential/commercial sector is also referred to as the built environment. In the United States, it consumed the largest amount of primary energy in 2005 (0.26 CMO). Energy is used for basic operations such as heating, ventilating, and air conditioning (HVAC); lighting; and operating many large and small appliances. In addition to the 39% of primary energy used for operations, about 12% more is consumed in making the materials required for building such as concrete, steel, glass, and aluminum. These building materials are among the most energy-intensive to manufacture. Together, the built environment consumes 51% of primary energy in the United States and represents a huge opportunity for realizing energy savings and for avoiding future energy expenditures.

The process for making cement requires heating lime and clays in kilns at temperatures exceeding 2,700°F (1,500°C). The process is very energy intensive, since every ton of cement consumes around 6 MBtu in production (which releases more than 1 ton of CO_2). Global production of cement in 2006 was about 2 billion tons, leading to about 5% of anthropogenic greenhouse gas emissions. As cement is an essential ingredient in many infrastructure projects (housing, roads, and bridges), its production is a strong indicator of economic development in the emerging economies. In 2006, China alone produced 1.2 billion tons, followed by India at 150 million tons, and the United States at 100 million tons. The production numbers are also a rough approximation of the consumption numbers, and the glaring difference in the production numbers for China and the rest of the world shows dramatically where most of the global construction is currently taking place. Production of concrete is the third largest contributor of greenhouse gas emissions after electrical power production and automobiles. Moreover, the process used today is essentially the same as that used 150 years ago. Since cement manufacture is responsible for about 5% of global energy use and greenhouse gas emissions, reducing energy requirements for concrete manufacture could have a significant impact on both energy and the environment.

The growth in residential/commercial energy use has paralleled the growth in the U.S. population. Some of the residential growth has occurred because more people have moved to single-family-dwelling suburban locations, but also because the size of dwellings has increased. The newer dwellings are built to higher standards of energy efficiency and have appliances that are also more energy efficient. Yet the overall trend is for American homes to require more energy per household, even as the number of occupants per unit decreases. The new refrigerators may be more efficient, but because they are larger they require more energy. Television sets based in liquid crystal displays are more efficient than their older cathode ray tube counterparts, but tend to be substantially larger in size. The newer home furnaces are more efficient, but they are heating larger homes. The average home size increased from 1,400 ft^2 in 1970 to about 2,500 ft^2 by 2006. These examples reiterate the fact that gains from improved efficiencies are often negated by the greater consumption that they encourage.

New building codes are resulting in higher building efficiency for residential and commercial structures. In the wake of the energy crisis of the 1970s, increased insulation in buildings also meant cutting back fresh air with the unintended consequence of building up the concentration of formaldehyde and other gases released from carpets and furnishings. The buildings developed what is referred to as "sick building syndrome." Technology has evolved since then. For example, ductwork that allows the out-flowing air to warm up the incoming make-up air goes a long way toward alleviating that problem.

Regulations that require appliance manufacturers to provide information on energy consumption help customers see the cost of operating them in addition to the cost of purchasing them. This practice has steered customers toward more efficient appliances. In the commercial sector, there has been an increasing desire for builders to use materials considered green, and more efficient HVAC and lighting systems, even if these are somewhat more expensive, because recognition is growing that their use adds value to the owners. The owners can in turn charge a premium for sale, lease, or rental of such properties. Even so, buildings last a long time and are replaced by new structures only at the rate of about 1–2% a year. For this reason, efficiency gains in the *already*-built sector are going to be slow.

A much larger opportunity in building exists in the developing countries. In India and China, it is estimated that the construction equivalent of the entire U.S. built environment will be taking place in the next 10 years—a business of several trillions of dollars per year. If energy-efficient measures are adopted there, net energy savings could amount to between 1 and 2 CMO. However, the greatest challenge will be delivering those energy-saving technologies at a price that the customers would be willing to pay. Take the case of Serious Materials, a company developing new building materials, such as windows, glass, and drywall, that have substantially reduced carbon footprints. One of their products, EcoRock, is a replacement for gypsum board sheetrock. Traditionally, production of sheetrock requires large amounts of heat to promote the reaction between the ingredients, and then for drying the product, calcium sulfate. EcoRock looks, feels, and behaves just like the standard sheetrock, but because it uses a chemical process that releases heat, the energy inputs for the reaction and drying the product are reduced by more than 80%. Kevin Surace, the CEO of Serious Materials, told the story that while he can make and sell this product at a price competitive with the standard gypsum drywalls (~$7.00 a sheet), when he tried to market his product in China, he found that the price of competitive sheetrock was less than $1.00, and his U.S. product was not wanted. If it could be made in China at $1.00 per sheet, it could be used in large quantity and substantially reduce the anticipated energy demands in the housing sector.

A substantial opportunity also exists when installing the lighting and appliances that are replaced several times during the life of a building. The motivation for replacing existing but not permanent fixtures and appliances with energy-efficient ones is different than when addressing energy efficiency of the building itself. Sometimes the incentives for making these changes are misplaced. For example, if a landlord is responsible for the utility bills, the tenant has less incentive for taking steps to conserve energy or to install more efficient fixtures. Nevertheless, one way that has proven quite effective for inducing people to replace inefficient fixtures with efficient ones has been the issuance of rebates by the utility. This practice is good from an energy efficiency standpoint, but not for the utilities as long as their income and profit rests on selling more energy.

California was one state that recognized this dilemma, and in order to get the utilities' to buy in, the state divorced the earnings of the utilities from the amount of electricity or gas they sold. This action allowed the utilities to engage in promoting energy efficiency. The utilities offered rebates for replacing older inefficient appliances with those having Energy Star ratings. Likewise, they promoted better insulation, hot water regulation, and other practices to their residential and commercial customers. The California Energy Commission, the body that oversees the utilities, boasts of the remarkable gains in efficiency realized by showing how the annual per capita electricity consumption in the state has stayed flat since 1973 (at 7,000 kWh, or 500 GO) while in the rest of the country it has increased 50% over this period, from 8,000 kWh to 12,000 kWh.

The rebates are issued by the electric or gas utility—but the costs of these rebates are included in the basic allowable costs that are used to determine the rates charged for their product. In other words, *all* of the utility customers pay for the higher base rates, but the individual purchaser of the efficient appliance or device gains the immediate benefit of the rebates.

In both residential and commercial buildings, providing adequate lighting comprises about 25% of energy consumption. Many observers have pointed out that an incandescent light bulb delivers only 3% of energy in the original fuel to useful light. This realization is leading to replacing incandescent lamps with compact fluorescent lamps (CFLs), which are about four times more efficient. The efficiency of lighting systems is conveniently expressed in lumens per watt (lm/W).[6] Whereas incandescent lamps have efficiencies between 12 and 15 lumens/watt, CFLs and other fluorescent tubes have efficiencies between 60 and 70 lm/W.

Using these more efficient lighting devices would reduce the energy requirements for lighting, and the associated carbon emissions by a corresponding amount. To assess how large of an impact they would make, let's calculate the savings from replacing one 100 W incandescent bulb with a 25 W CFL. If the light bulb were used for 6 h each day, the savings would amount to 75×6 Wh or 450 Wh, and over a year the savings in energy would amount to 164 kWh, for a savings of about $20. If a household had a dozen such lamps, the dollar value of savings would provide a significant incentive. Now let's ask the question of its impact on overall energy consumption. How many such light bulbs do we need to replace in order to have an annual savings of 1 CMO? The answer is a hundred billion bulbs! That is more than 15 bulbs for every man, woman, and child on the planet. Clearly, while switching to CFLs does ease our energy requirements, its potential impact on a global energy scale will be rather limited. Again, it does not mean that we should not switch incandescent lamps to CFLs, but we must do so with the realization that by itself it makes only a small contribution to reducing energy demand. As we examine the various opportunities for shaving off energy use, we see that there are limited prospects of energy savings on the CMO scale.

Aesthetic considerations have also come into play to limit the penetration of CFLs in the market. People often find the quality of light from CFLs less pleasing, and since they generally cannot be used with dimmers, they offer less functionality than do incandescent bulbs. The importance of aesthetics can be

---

6. Comparison based on lumens instead of light energy is desirable, because a lumen also takes into account the response of human eye to light of different wavelengths (color).

paramount in certain business settings. Retailers tend to use track spotlights with halogen bulbs instead of CFLs, despite the former's high cost of installation and operation, because when customers look at themselves in the mirror to see whether the product suits them, they should look their best; not ashen and sickly as CFLs tend to make them appear, which could ruin a potential sale.

In the commercial sector, particularly in factories, warehouses, streets and parking lots, much of the lighting is provided by high-intensity discharge lamps. These lamps have efficiencies exceeding 150 lm/W, considerably more than fluorescent tube fixtures. Most of the light that we see when flying over cities at nighttime (as shown on the cover of this book) is from these high-intensity discharge lamps. Reducing power consumption in these devices either by directing the light toward where it is needed, instead of letting it scatter upward, or by building the fixtures with sensors so that they can be dimmed (or turned off) when people are not around, could have a substantial impact on energy use in the built environment sector.

There are additional lighting products based on light-emitting diodes (LEDs). They have efficiencies between 100 and 150 lm/W. Fixtures based on LEDs have many desirable attributes, such as a full range of colors and long lifetimes. Some of them also come as large, flat illuminating surfaces instead of bright point objects, and thus offer architects a completely new way of lighting. Currently, LED-based fixtures are considerably more expensive than other lighting systems, and they are generally not suited as a direct replacement of conventional fixtures. For these reasons, they are likely to require more cycles of innovation and price reduction before they are widely adopted.

The U.S. Department of Energy is currently developing near-zero-energy consuming homes, which would cut a large amount of the energy consumption from this area if widely used. Even without this measure, the Lawrence Berkeley National Laboratory calculates that in the U.S. savings of about 0.10 CMO are possible in residential buildings between 2010 and 2030 with simple increases in the efficiency of appliances, demonstrating the immense potential for savings that is available. This figure is even exceeded by the 16.8 quads (0.11 CMO) in efficiency savings available from commercial and industrial buildings combined, which would use many of the same measures as in the residential sector. Now, if we want truly substantial savings in the built environment in the United States, these can be had using conservation measures such as a shift away from the so-called McMansions to high-density dwellings. Innovations that make those dwellings desirable to inhabit are sorely needed. They could provide items that minimize things people dislike about high-density housing, such as noise and traffic, or they could be geared toward enhancing the quality of life in a city such as communal gardening or walking access to local markets, cafes, and other places of interest.

The current interest in conservation and environmental protection has people trying to reduce their carbon footprint. And there are many opportunities that reduce overall energy consumption, save costs, and reduce CO_2 emissions. Many Web sites allow people to estimate their carbon footprint by answering questions about their lifestyle. There are also companies promoting products that monitor energy use throughout the home and list on one screen how much electricity is being used and where. The customer can use that information and turn off unnecessary appliances or lights and receive instant gratification as the screen now displays the figure for the reduced energy consumption. The direct energy savings from these information products may not be substantial, but the greatest value lies in their ability to change our habits—transforming us from being profligate users of energy into frugal consumers. Cumulative energy savings from resulting *conservation* would be significant.

Some companies also offer the opportunity to buy carbon offsets and pledge to use the funds to plant trees or engage in other actions that would mitigate the impact of the person's actions. While we don't deny that increased awareness of the consequences of one's behavior is sure to have some impact on the behavior, we doubt if we can buy our way out of the energy crisis—buying carbon offsets is so reminiscent of buying indulgences for one's sins. If we are to make a substantial impact on greenhouse gas emissions, we have to examine our food habits. It is unfortunate that energy compilations do not list food as a distinct end-use category, but lump its components in the industrial, commercial/residential, and transportation sectors. The impact from reducing meat and fish in one's diet can be much greater than the efficiency-improving technologies we discussed.

Transportation

Transportation consumes about a quarter of the primary energy and plays an important role in the energy demands of the world. Transportation is the nexus where issues around energy supply, climate change, and energy security collide. As mentioned above, the transportation sector overwhelmingly depends on petroleum, deriving more than 90% of its primary energy from this single source.[7] Concerns about Hubbert's peak (see chapter 5) most directly affect transportation, because more than 90% of the energy for transportation comes from one source—petroleum. Some alternatives, such as coal-derived liquid fuels, or electric cars running on coal-derived power, relieve the demand on imported petroleum but have the potential to markedly increase greenhouse gas emissions.

The role of transportation is nowhere as significant as in the United States with its vast land mass and large population—now the third larg-

---

7. Although some of the petroleum is used as feedstock for chemicals and plastics, the impact of dwindling petroleum is not likely to be felt on those commodities as much because they tend to command higher prices and the cost of petroleum represents only a part, not the totality, of the price.

est in the world. The nation's use of automobiles has been strong from the time of its invention in the late 19th century and the development of mass production techniques in the early 20th century that made the automobile affordable for many. From approximately 4,000 vehicles produced in the United States in 1900, the annual demand has grown such that 17 million vehicles were put on the road in 2006, including automobiles, sport utility vehicles (SUVs), light trucks, buses, and heavy trucks. Sales and leases of vehicles for personal transport accounted for almost 70% of all new vehicles. Light trucks and SUVs accounted for more than 50% of the vehicles used for personal transport. Because these large vehicles generally have poorer fuel economy than the usual automobile, their popularity has had a significant impact on the use of automotive fuel. Following the spike in oil prices in 2008, there has been renewed interest in the public for fuel-efficient vehicles.

An important factor in the use of automotive fuel in the United States since 1975 has been the application of Corporate Average Fuel Economy, or CAFE, regulations. In the wake of the oil embargo of 1973–1974, the U.S. Congress enacted a law directed at improving fuel efficiency of automobiles and light trucks weighing less than 8,500 lb. The average fuel efficiency for automobiles in 1974 was a measly 13.5 mpg. The National Highway Traffic Safety Administration (NHTSA) was charged with the responsibility for administering the CAFE program with the near-term goal of doubling fuel efficiency by 1985, to 27.5 mpg, while considering factors such as technical feasibility, economic practicality, effects of pollution control standards on fuel economy, and the need for the nation to conserve oil.

Under the CAFE regulation, each manufacturer must produce a fleet of cars that, on average, meets certain mileage performance standards established by NHTSA. Penalties for not meeting the established CAFE standards was set at $5.00 per tenth of mile per gallon under the set target, multiplied by the number of vehicles sold in that model year. According to NHTSA, since 1983 automakers have paid penalties of more than $500 million. Asian and domestic automakers have never paid any penalty, but many of the European manufacturers, whose sales in the United States include mostly high-performance cars, have regularly paid millions in penalty.

Since 1985, though, the CAFE standard has not changed. Increasing CAFE standards have met with stiff opposition from auto manufacturers, who say that any further increases would hurt their profitability and risk their viability. It is worth noting that the actual average fuel economy of the vehicles in the United States is significantly lower than the CAFE standard because the standard is measured under somewhat unrealistic driving conditions (no speeding, no climbing of hills, no acceleration, no use of air conditioners, etc.). The on-road average fuel economy data for

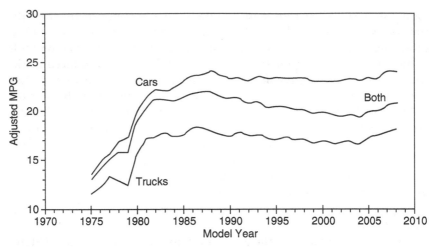

Exhibit 8.5. On-road Fuel Economy of U.S. Vehicles (Source: EPA)

vehicles in the United States have been compiled by the EPA, and their data are displayed in exhibit 8.5.[8]

What jumps out is the surge in fuel economy (measured in miles per gallon) that occurred between 1975 and 1980, after which the gains slowed down and have been essentially flat since 1985 for cars, when oil prices had fallen from their historic highs and the political will had changed. Several factors account for the tailing off of fuel economy. First, as mentioned above, the CAFE standard has not changed since 1985. However, efficiency gains in engine design have continued to occur, and modern engines deliver more motive work, which is measured in brake horsepower-hour, per gallon of fuel. Those gains were offset by the increases in the fuel-consuming performance of the vehicles (faster acceleration, expressed by average time for accelerating from 0 to 60 mph, which decreased from 13.5 s to 9.5 s over this period) and the increased weight of the vehicles (average vehicle weight increased from 3,300 lb to 4,200 lb). Since 1985 the sales for light trucks (including SUVs) increased dramatically, capturing a much larger share of market (to 30%, from less than 10% in 1985), and this shift, which reflects a preference for heavier SUVs and high-performance cars, is a major factor in the stagnation of the fuel economy. Whether this preference stems from safety considerations or other factors, it provides *partial* justification for the automakers claim that increasing CAFE standards would seriously affect their bottom line.

8. Fuel economy rated by the EPA was historically based on driving on level ground with no acceleration and no air conditioning. The actual fuel economy realized by owners is about 15% lower than the EPA rating. Even though the EPA has since changed its protocol, the CAFE standards are still based on the original EPA protocol.

In the past, to stem fuel consumption, many European countries used high fuel taxes, several times those in the United States. They also levied vehicle taxes based on both motor horsepower and vehicle weight. Now Europe is moving toward a standard based on CO_2 emissions.

The CAFE standards and fuel economy are generally quoted in miles per gallon. As Larrick and Soll pointed out in their 2008 *Science* article,[9] a fixation on highest "miles per gallon" can lead to an underestimation of the benefit—net savings of fuel—from replacing grossly inefficient vehicles with those that are even only marginally more efficient. They illustrate the point with a choice that a family might consider: replacing one of their two vehicles with a more efficient one. The family owns an SUV that gives 12 mpg and a sedan that gives 28 mpg. Both vehicles are driven 10,000 miles a year. They could replace the sedan with a more efficient sedan that gives 40 mpg (a 43% increase) or their SUV with a new SUV that gives 14 mpg (a 17% increase). Contrary to the expectation that switching to a more efficient sedan would lead to greater savings, the amount of gasoline saved upon replacing the SUV is actually larger. The SUV with 12 mpg consumes 833 gallons a year (10,000 mi/12 mpg), and the one with 14 mpg consumes 714 gallons, a net savings of almost 120 gallons. Because their current sedan is already quite efficient, it consumes 357 gallons (10,000 mi/28 mpg). The replacement sedan would consume 250 gallons and save the family only 107 gallons a year. Thus, if the objective is to minimize fuel consumption, it is not sufficient to simply look at the respective mpg ratings; we must consider the actual miles the vehicles would be driven and calculate the net fuel savings.

Deployment of gas-electric hybrid engines is one way of achieving higher fuel efficiencies. Toyota has been very successful with its Prius gas-electric hybrid, which has an EPA fuel economy of 48 mpg. This car uses a parallel hybrid system; it has a modest four-cylinder gas engine and a relatively small battery pack, weighing 60 lb and consisting of 168 nickel-metal hydride (NiMH) cells, with a total capacity of 1.4 kWh that would suffice for about 5 miles if used alone. The electric system in this car supplements power to the gas engine during start-up and accelerations, but once the car has achieved a cruising speed, the gas engine takes over. During deceleration and braking, the kinetic energy of the vehicle is used to charge up the battery. The relatively small gas engine and the electric motors work in parallel, and the car's on-board computer seamlessly draws on these resources depending on the driving conditions and the level of charge in the battery.

Beyond the parallel hybrid designs used thus far by Toyota and Honda, some observers point out that series-hybrid designs can offer significant

9. R. P. Larrick and J. B. Soll. "Economics. The MPG illusion," *Science,* vol. 320, p. 1593–1954, 2008.

further increases in efficiency and also cost savings from elimination of the mechanical drive train. In this concept, an internal combustion engine operated at or near its optimum speed generates electric power, most of which is sent directly to high-efficiency induction motors mounted on the drive wheels, while a small amount is sent to a modestly sized battery pack for use in acceleration. The drive shaft and transmission are eliminated, providing savings in both weight and cost.

Apart from hybridization, there are a number of technologies that could be implemented today to increase the fuel efficiency of vehicles. Variable speed transmission is one technology whose benefits to fuel economy have been pointed out by many researchers, including a panel of the National Academy of Sciences. Reducing the weight of the vehicle is another obvious candidate for improvements. Think about it: using a 3-ton vehicle to move a 150 lb person has to be one of the most inefficient ways to transport people, not to say "stupid," when the world energy supply is short and our living space is degraded by using it!

At present less than 1% of fuel goes to moving the driver; the rest goes to either transporting the car (~12%) or to heat loss waste and noise (87%). We have implicitly come to believe that a heavier vehicle is a safer vehicle, and this mentality has contributed to the steady escalation in the weight of vehicles. That attitude, though, is beginning to shift as more and more small cars are getting high marks for safety on tests run by the Insurance Institute of Highway Safety. Amory Lovins and associates at the Rocky Mountain Institute have shown that many light-weight composite materials are indeed as strong if not stronger than their counterparts currently used in the industry.[10] They point out that much of the increased fuel efficiency of hybrid cars arises because of their lighter weight, not from the hybridization of the engine. They have also developed designs for a five-passenger sedan that weighs around 1,900 lb and can go 66 miles on a gallon.

To tap into the creative genius of the public at large, the X-Prize Foundation has announced a $10 million prize for developing superefficient cars with a target fuel economy of 100 mpg. The cars have to be designed for everyday use and be ready for manufacturing. We can hope that many innovative engineers will step forward and take up the challenge. Following the global financial crisis in 2008, the auto industry has been hurting. The potential upside of this downturn is that now may be the time for innovative concepts to take hold.

The weight reduction of cars can be accomplished by using lighter steels, replacing some of the steel components with aluminum, or using components made with carbon composites. Carbon composite materials are more expensive by the pound, but fewer pounds of this material are

10. A. Lovins and E. K. Datta, *Winning the Oil Endgame*, Rocky Mountain Institute, Boulder, Colo., 2005.

needed than when steel is used. Further cost savings can be realized because of their simplified manufacturing. According to the Rocky Mountain Institute, the increased cost for using lighter materials would be recovered from the fuel savings within a couple years worth of driving. According to some other estimates, fuel economy of vehicles could be enhanced 25% by incorporating improvements costing no more than $1,000. And then there are other measures that can easily gain 10% efficiency and cost nothing except some vigilance on our part such as maintaining proper tire pressure, reducing the cruising speed to no more than 65 mph, and avoiding fast accelerations.

Use of electric cars as a strategy for displacing oil has been gaining traction lately. In the 1990s, General Motors had introduced the EV1, an all-electric car, in response to California's mandate for zero-emission vehicles. When the state withdrew that requirement, GM recalled the cars and had them crushed. Much has been written about the factors leading to the demise of the EV1. The documentary "Who Killed the Electric Car" finds many parties that could be blamed for it: the auto industry, big oil, battery technology, an overzealous California Air Resources Board, and the public at large, to name some. The fundamental technical problem with EV1 and early electric cars had to do with the poor energy density of their batteries. Lead acid batteries have an energy density of around 0.1 MJ/kg, which means that it takes 78 lb of battery to hold 1 kWh of charge—the energy sufficient for less than 4 miles. Compare those 78 lb of battery with a gasoline-powered car that could go 4 miles on a fifth of a gallon that weighs about one and a quarter pounds. The total weight of batteries alone in a first-generation EV1 car was about 1,500 lb. Of course, batteries are not consumed and gasoline is, but still the disparity in the energy density on a weight basis placed a severe fuel penalty on the electric car even when allowance was made for the approximately 400 lb heavier engine block in a gas-powered vehicle compared to about 100 lb for electric motors.

The second-generation EV1 came with NiMH batteries, which have about twice the energy density, and therefore weigh half as much, as lead acid batteries. This helped with the weight issue but these vehicles had their own set of problems, including that their batteries were expensive and could not be cycled more than about 500 times. Compared to lead acid batteries, which sell for $170/kWh, NiMH batteries cost about $1000/kWh. On a full charge of 26.4 kWh, the car could go 160 miles. Since 26.4 kWh of electricity from the grid would cost about $3.00, the operating expense (2¢/mile) was very low and compared favorably even with gasoline at $1.00/gal. But, since the NiMH batteries could last only about 500 cycles of charging and discharging, amortizing the cost of the batteries ($26,400) over the 80,000 miles battery lifetime gave a cost per mile increase of 33 ¢, more than ten times the cost of grid electricity for powering the car. The limited driving range and the total cost of batteries remain the principal barriers to broader adoption of electric vehicles.

As mentioned above, Toyota's Prius also uses NiMH batteries, albeit of a considerably smaller capacity (1.4 kWh vs. 26.4 kWh for the EV1). This strategy has allowed Toyota to control the overall production cost, and it can manage many more charge-discharge cycles by making sure that the batteries are not discharged deeply.[11] If the NiMH batteries are kept within 60% and 90% of the full charge, they can have significantly longer lives, but a deep discharge would be required for electric vehicles (or for plug-in hybrids) to keep the weight of the battery pack under one ton.

Lithium-ion (Li-ion) batteries have about 30% more energy density than NiMH batteries, and they can also be charged relatively rapidly. There are several varieties of Li-ion batteries; those based on cobalt oxide have high energy density, while lithium iron phosphate batteries have high power density, which is good for acceleration and fast charging, but they sacrifice energy density and thus reduce the range. Lithium ion batteries also cost more, about $2500/kWh. This high cost is not a major factor in applications like laptop computers and cell phones, but in a car with a driving range of, say, 200 miles, the batteries could cost about $125,000. If the battery lasts 1,000 cycles, the cost per mile for the battery alone would amount to more than 60 ¢/mile. Ignoring the relatively minor cost of grid electricity, the 60 ¢/mile cost would be equivalent to gasoline selling for $12/gallon. Reducing the cost of batteries is a key R&D area, and these costs are being driven down.

In view of the high cost of batteries and fact that electric cars with Li-ion technology can deliver high performance, several electric car start-ups, notably Tesla and Fisker, have directed their first products at the high end of the market. The first customers of these cars are not going for economy; they are likely motivated by the statement that driving a high-tech, high-performance electric car would make.

By focusing on this select customer base, Tesla and Fisker hope to start drawing in revenues to develop more affordable models. Their plans have spurred the traditional automakers into also announcing their plans to introduce (or reintroduce) electric cars. All of them are initially going to be quite expensive. Some analysts project that with the advances in battery technologies, we can expect hybrid, plug-in hybrid, and all-electric vehicles to cost no more than $2,500, $5,000, and $14,000 more than a comparable gasoline car respectively. There is increasing expectation that many more all-electric or plug-in hybrid cars will be on streets around the world. Their proliferation will open many other business opportunities.

Many companies are looking at setting up charging stations in parking lots or hotels and other places where people who are away from home may be

---

11. In models sold in Europe and Japan, Toyota offers a plug-in option that allows drivers to choose an all-electric mode. The car primarily runs on electric power with the expectation that before the battery is fully discharged, the car will be back at a place (home) where it can be recharged.

able to charge their vehicles. As charging of batteries can take several hours, a start-up company—Better Place—is developing a business around cars with swappable batteries. They are working with auto manufacturers to design their electric vehicles with the potential for quick replacement of batteries. Their business model is based on providing customers a means of quickly replacing drained batteries with precharged ones at service stations in the time it takes to refill the fuel tank. Since the urgency of recharging the battery when the customer is waiting is removed, Better Place can work with electric utilities and charge up the batteries at times of low demand, as well as provide backup to the utilities from charged batteries at times of high demand. As the owner of the batteries, Better Place also relieves the car owners of the concerns over battery life. The concept is not without its challenges. As mentioned above, batteries for electric and plug-in hybrid vehicles weigh almost a ton. For reasons of better handling and maneuverability, car designers would prefer to distribute the weight and that may not lend itself to a quick and easy swapping of batteries.

These developments in reducing price and increased convenience would increase the penetration of electric cars in the market, as would also the development of a smart electric grid that could use the parked cars connected to it as a buffer for load shifting—charging or discharging them as the demand for electricity changes during the day. It will be like the question of the chicken and the egg: unless an appropriate infrastructure is in place, markets for these sophisticated electric cars are not likely to proliferate, while without a large market for these cars, the incentive for investing in the necessary infrastructure is lacking. In anticipation of a rising demand for electric and plug-in hybrid cars, some businesses are developing sound systems for electric cars to alert pedestrians and others of the otherwise very quiet electric cars as they approach. Still others are offering to convert existing cars into electric vehicles.

This book is all about energy, and the question we need to address is what the ramifications are of substantial electric transportation to the total energy supply. Any use of electric transportation will certainly reduce the demand for petroleum. But for that reduction to become significant, say, even at a level of 0.01 CMO, the number of electric cars has to sizable. Let's assume we are replacing high-performance cars having relatively poor mileage, say 10 mpg, that are driven 10,000 miles a year. On average, these cars would consume 1,000 gallons of gasoline, and it would take 10 million electric cars to save 0.01 CMO. Note that it took more than 10 years for Toyota to sell its millionth Prius, which suggests that saving 0.01 CMO/yr by this approach, would take decades.

We must also ask whether there are sufficient resources of lithium to support an electrification of hundreds of millions of vehicles. The main sources of lithium today are dried salt lake beds in Argentina, Chile, and Bolivia. Another large resource of lithium is in Tibet. Currently, use of Li-ion batteries is mostly for cell phones, PDAs, and laptop computers. Estimates of lithium carbonate reserves are between 100,000 and 200,000 tons. A Li-ion battery for an automobile application would require several hundred pounds of lithium per vehicle, so global reserves of lithium would be exhausted by the first couple million vehicles. There are other sources of lithium, but extraction of lithium from them requires a very different and considerably more expensive technology. As with other minerals—including oil—lithium *reserves* depend on the price and technology. Other technologies for extracting lithium would have to be developed if electric vehicles using lithium-ion batteries gain a significant market share of the automobile sector. It would be ironic if producing the "green" electric car ended up damaging ecosystems!

But let's suppose that electric cars become wildly popular, lithium is available, and half the world's automobile fleet is either all-electric or plug-in hybrid. In such a scenario, we will need to increase our electric power production from the current 1.2 CMO to about 1.6 CMO to power these vehicles. A 2007 study by the Pacific Northwest National Laboratory[12] made the statement that there is sufficient capacity in the existing electric grid to support conversion of half of the cars in the United States into electric vehicles. This finding has been misrepresented in the media to suggest that we would not need extra electrical energy. Instead, the Pacific Northwest study makes the point that new power plants might not be needed if the cars were charged overnight during nonpeak hours, but we would still need to produce more energy, mostly by burning more coal, to offset the energy the cars would have used from oil. If, on the other hand, people plugged in their vehicles immediately upon returning home from work, the grid would be severely stressed, as this timing would coincide with the summer evening peak demand, when offices, commuter trains, and home HVAC systems all seem to be in use. The impact on CO_2 emissions would depend on whether the extra 0.4 CMO for electric power was met by burning coal or from a carbon-neutral source.

Moving freight is another significant component of the transportation sector. According to the U.S. Bureau of Transportation Statistics, about 43 million tons of goods are transported 280 miles daily for a total of 12 billion ton-miles. Shipping by barges along waterways is most energy efficient, although limited to locations connected by waterways. Railways are the next most energy efficient way to transport freight, and trucks are the least efficient. Yet, because of the convenient door-to-door service that trucks can provide, in the United States they carry about five times the tonnage shipped by rail. Because the distance of shipment by rail tends to be significantly longer, on a ton-mile basis, the two modes are comparable. To the extent that more of the freight can be switched from trucks to trains or ships, we can expect a reduction in energy consumption in this sector.

12. M. C. W. Kintner-Meyer, K. P. Schneider, and R. G. Pratt. 2007. "Impacts Assessment of Plug-in Hybrid Vehicles on Electric Utilities and Regional U.S. Power Grids: Part 1: Technical Analysis." Online Journal of EUEC 1:paper # 04.

In view of the difficulties in increasing petroleum supplies by even a few percent, on the one hand, and producing sufficient quantities of biofuels on the other, plus the challenges of increasing the penetration of electric cars, it becomes clear that raising CAFE standards offers the quickest and most feasible way to save on petroleum consumption.

Lifestyle Changes

There is a real opportunity for conservation based on a significant change in living patterns. Based on the different scenarios, we estimate possible demand reduction through conservation of as much as 3 CMO/yr by the year 2050 from the projected business-as-usual scenario of 9 CMO/yr. For this to occur, citizens will have to be joined by city and regional planners and financing agents—including private financing institutions backed by local, state, and federal governments.

The first step in this long-term, extensive, and expensive process would be a return to more closely linked living arrangements similar to those that were used before the universal use of the automobile, including perhaps a greater use of apartment living. Apartments are inherently more energy efficient because they have fewer walls exposed to the extremes of heat and cold characteristic of much of North America and many other parts of the world. The desire for bigger houses, multiple automobiles, and a house in the suburbs has resulted in greater per capita energy use that, compounded by a substantial population growth, has led to ever increasing demand for energy in these United States. Andrew Bacevich noted in his 2008 book *The Limits of Power* that the current tendency of American citizens toward acquisition of "stuff" is not a recent phenomenon; it was also noted by Alexis de Tocqueville in the 1830s.[13]

The second but essentially simultaneous action would be a radical overhaul of the urban design to one based on high-density, but desirable, multiuse developments. Historically, the urban-suburban interface was developed ad hoc, and any planning that occurred did not include consideration of energy savings. New developments should include this factor if we are to conserve. Below we suggest a new version, most likely a reconfiguration of our now popular suburban living environment. It will surely require much thought, and investments are so large that only the federal and state governments could together finance such an endeavor. Urban population in the developing countries has been increasing sharply, and it offers an opportunity to develop new megacities in a way that encourages conservation.

Significant reduction in our energy consumption for transportation is possible through the use of public transport or ride sharing. In the United

13. Andrew J. Bacevich. *The Limits of Power: The End of American Exceptionalism,* Henry Holt, New York, 2008.

States, except for a few metropolitan areas in the East, public transport is woefully inadequate. There are many historic reasons why the cities evolved in this manner, but absent convenient and reliable public transport, there are few options for people other than using their cars. Oftentimes it is the difficulty of bridging the proverbial last mile that forces people to travel by car. Measures that would encourage large numbers of people to share rides or take public transportation can have a much larger impact than improving car efficiencies by a few percent. Many businesses are seeing opportunities in this arena, introducing services that make use of cell phones and the Internet to connect potential riders with rides. These actions could lead to significant conservation, but again, they require changes in personal attitudes.

What if we were to construct a series of relatively small hamlets linked by a common transit line that connected them to reach a central city, and/or a major industrial/academic complex? The hamlets would be equipped with suitable apartments. Preferably these buildings would have small shops serving local needs for small consumer goods and activities such as vegetables, meats, and general small household and other merchandise found in supermarkets, as well as dry cleaning and hairdressing establishments and other shops of a similar nature. These small hamlets would be surrounded by homes that could accommodate larger families or other people who wished extra space and would be willing to pay well above the average for the privilege—perhaps their taxes would be set at a higher rate, too—but in any case they would be sited within distances from the hamlets accessible with bicycles or golf-cart-like conveyances.

Within a few miles, or transit stops, the hamlet citizens would find a larger variety of shops and conveniences—movie theaters, for example, if the latter were not outmoded by then. The lines would extend to major cities that would, most likely, be smaller than those of today but still the center of theaters, museums, concert halls, and other activities requiring larger audiences to be viable. Contact between the larger cities could be maintained by highways, or supplemental rail transit links, if the traffic justifies.

Naturally, the buildings in the new hamlets would be designed to take advantage of the passive solar energy for heating and lighting. They would be built to meet the highest efficiency standards and sited to benefit from sunshine orientation. They would be equipped with the most efficient appliances. In many cases supplies from several diverse sources, including income sources, would meet their electricity and gas needs. In some cases, the hamlets could be configured to benefit from the waste heat from nearby power production facilities that use inherited fuels. The low-pressure steam or hot water from these plants could be supplied to nearby dwellings. Such schemes are in active use in the Nordic countries and Russia, for example, although limited in their effective range by cooling of the steam.

Substantial reduction in energy demand could also be achieved by connecting between major cities by freight rail and their dependent satellites by roads. This would entail remaking of a national railroad complex that would consist of a coordinated long-haul rail and short-haul truck transport, in most cases less than 100 to 200 miles. East-west and north-south highways would be located to provide for passenger auto, not freight transit.[14] The lighter loads imposed on the highway by cars and the occasional truck would make highway maintenance a much smaller burden on the taxpayer.

Another behavior change that can have a substantial impact on overall energy use is shifting one's diet toward vegetarianism. We mentioned a study by Eshel and Martin in chapter 3, which analyzed energy inputs required for producing a range of diets.[15] Because production of meat takes many times its weight of grains and other meats, diets that derive a higher proportion of their caloric content from meats can require energy inputs that are two to three times as much as vegetarian diets. The impact of switching to vegetarian diets on greenhouse gas emissions is even greater, and as we discussed in chapter 3, they are equivalent to avoiding 300 GO/yr per capita. Opportunities for individuals to reduce greenhouse gas emissions on this large a scale are rare.

The impact on total energy from the various conservation measures mentioned here could be substantial if they are widely adopted. However, since the prospects for people in large numbers relocating to high-density living arrangements, favoring mass transit, or switching diets are not very high, it would be not be prudent to count on these for making a difference on the scale of a CMO.

Summary

Conservation and efficiency improvements offer the quickest and easiest ways to save significant quantities of energy. Efficiency improvements have been occurring all along, but increased awareness about dual challenges of energy shortage and greenhouse gas emissions could motivate further technical advances and a wider adoption of those practices. Many of the measures do not require any sacrifice and also provide savings in cost. These measures include turning off unneeded lights and appliances, switching to more efficient appliances, and installing better insulation. There are also other changes

14. When President Eisenhower conceived the national interstate system, he was not anticipating the extent of its use as the major conduit for transporting people and goods. He was approaching it from a strategic standpoint. Unlike railroads that, if destroyed, could not be used unless fully repaired, he had witnessed how highways could be easily bypassed and/or easily patched as he commanded the Allied Forces in WW II. It made sense to connect the nation with a system of roadways in preparation for future conflicts with the then aggressive Soviet Union and its nuclear weapons capability.

15. G. Eshel and P. A. Martin, "Diet, energy, and global warming," *Earth Interactions,* vol. 10, pp. 1–17, 2006. DOI: 10.1175/EI167.1.

that if implemented would reduce energy consumption, but their cost may take several years to recover.

Conservation measures are key because by slowing down the anticipated increase in energy demand they buy us time to develop (a) our income resources to the point that they become affordable for larger scale adoption, and (b) our untapped inherited resources as necessary. Government regulations like CAFE standards and Energy Star ratings help promote energy efficiency, but these prescriptive measures must also be matched by responsible behavior on the part of customers, or the potential energy savings will not be realized.

9

The Path Forward

In this book we reviewed the course of energy consumption over the ages and projected the level of consumption through 2050. To facilitate the discussion, we introduced a new unit of energy—a cubic mile of oil equivalent, or CMO—that enables description of global energy flows in terms and numbers that are immediately comprehensible. We surveyed the various sources of energy in current use, established the quantities used, and projected our future needs on a global basis.

While for much of our history the availability of energy has played an important role in determining the potentials and abilities of humans, in recent times energy has become much more important because resources are coming under strain. Greater energy use is beginning to influence our environment more strongly than ever before. A characteristic of global energy supply systems is the slowness with which they can shift. The slowness is a consequence of several factors. The size of the incumbent technologies and the advantage that they have in terms of learned improvements, economies of scale, and delivery infrastructure, play an important role in the rate at which new technologies are adopted.

New technologies are often more expensive simply because cost reductions occur with experience, and it takes time for something new to penetrate the markets and build an experience base. Government subsidies and research and development (R&D) investments can help break this vicious cycle, but in the end the technologies have to deliver value to the customers before they can be adopted widely. There are also limited numbers of manufacturing and delivery systems in place for new technologies. Because basic energy supplies adapt slowly to change—as do most technologies—while energy demand grows more rapidly with population and income, we must act now to bring new supplies and new patterns of energy demand into play to meet the projected global energy demands of mid 21st century. *Time is of the essence!*

Abundant energy has become an essential part of modern life, and we cannot go without it if we wish to retain even a small fraction of our current civilization. While as hunters and gatherers humans subsisted on per capita annual energy consumption of about 50 GO (gallons of oil equivalent), and as settled agrarians our energy consumption was about 120 GO/yr, modern society requires about 1,000 GO/yr per person, which *if* made available to 6.5 billion people around the world today would place global energy consumption at 6.0 CMO/yr right now. Energy should be supplied to all citizens and nations of the world in an equitable and affordable manner. Increasing world populations, expected to grow at almost 1%/yr for the next decade or more, and increasing wealth expected to grow at 2.5–3%/yr, combine to create an increased demand for energy. Two of the world's largest nations—China and India—with nearly 40% of world's population, have in recent years experienced incomes growing at rates of around 10%/yr and almost 6%/yr, respectively. These growth rates are far above the global average, but many other countries in Asia, South America, and sub-Saharan Africa also have economies growing vigorously with inflation-adjusted annual GDP increases of greater than 4%. The increased wealth naturally leads to increasing demand for goods and services requiring more energy.

As discussed in chapter 8, some developed nations are able to increase their per capita GDP without a concomitant increase in per capita energy consumption because of the gradual shift from manufacturing to nonmanufacturing economies. In contrast, the emerging economies must be concerned with building roadways, railways, waterways, and sewers, laying down an electric power grid, and providing an information superhighway to (a) provide a better and healthier environment for their citizens and (b) enable them to fully participate in a global economy. This development is also an opportunity for all to sell their goods and services in a vastly expanded global economy.

The world currently consumes 3 CMO of energy annually. It uses 1.0 CMO from oil, 0.8 CMO from coal, 0.6 CMO from natural gas, and approximately 0.2 CMO each from hydropower, nuclear, and wood.[1] Although its population is only 1/20th of the world population, the United States uses about one-fifth of the world's energy. The drivers for substantial increases in global energy demand are in place: more than three billion people are poised to sharply increase their standard of living, and in India and China there are already large groups of people whose wealth equals that of the average citizen of the more affluent countries such as Sweden and Switzerland. Business as usual for the world—which *includes* a steady improvement in efficiencies as witnessed

---

1. The values in this chapter are the same as those presented in preceding chapters, only rounded.

during the past 40 years—would place the annual global demand for energy in 2050 at around 9 CMO. Even if we were to follow a more modest growth scenario, the annual global energy demand could still increase to 6 CMO by then. We therefore project a need for energy sources capable of delivering a minimum of an additional 3 CMO annually. Fifty years from now, the 1 CMO a year we now obtain from conventional oil will also have to be replaced, adding another CMO to the demand for alternative sources. And if we wish to reduce the role of coal and natural gas, then the alternative sources will need to provide a total of between 4 and 5 CMO by 2050.

Meeting this energy demand by mid century will not be easy and will inevitably require significant changes in the world's patterns of production and consumption of energy. The fact that an overwhelming number of energy users, namely us, lack knowledge of where it comes from and its attributes—including its effect on the environment and their future—will be an important obstacle to making truly informed choices, plans, and key changes in our energy production and use. Remedying this deficiency through educating ourselves and engaging in general discussion of the issues is vital.

We must collectively address the question of where are we going to get the extra 4 to 5 CMO of energy annually to meet the projected growth? We have to know what it takes to produce even 1 CMO of energy from each of the different sources in order to weigh our options and make the appropriate trade-offs. For this analysis, it is important to recall that all forms of energy are not equally useful and, therefore, not directly comparable. Electricity is one of the most versatile forms of energy, and 40% of world's primary energy goes into producing electricity before the energy is consumed by the users. Furthermore, many alternative sources of primary energy, such as nuclear, hydroelectric, and photovoltaics (PV), directly produce electricity. We therefore use electricity as the "currency" of energy and treat the production of 1 kWh of electricity as being equivalent to 10,000 Btu/kWh, which is about average for the current fleet of thermal to electrical power plants. At this discounted rate for thermal sources of energy, 1 CMO is effectively equivalent to the 15.3 trillion kWh that can be generated from it at approximately 34% efficiency. This amount of electrical energy could be generated annually from an installed capacity of 1.7 TW assuming the source produces energy 24 hours a day and 365 days a year. To the extent that power generation sources have some downtime due to lack of availability, maintenance, or other reasons, the required capacity would have to be correspondingly larger than 1.7 TW. Installed capacities for solar and wind facilities, which are typically available a quarter to a third of the time, would have to be rated between 5 and 6 TW to produce 1 CMO annually.

Producing 1 CMO per Year from Various Sources

How, then, shall we consider meeting this increased demand of 4–5 CMO/yr? Can we place our hopes on an unprecedented growth in the use of one or more of the income resources? We summarize here what we have discussed in preceding chapters about different energy sources in terms of the resource potential, and also what we would need to do in order to build sufficient capacity to generate 1 CMO of energy annually from each of the various sources by 2060.

Petroleum

The world has about 46 CMO of conventional petroleum reserves. Recall that estimates of reserves and resources depend on the prevailing price and the state of technology and are therefore subject to change with time. The current resource base of additional conventional petroleum is much larger than the reserves, estimated between 35 CMO (likely) and 94 CMO (speculative). If oil consumption keeps increasing at 2%/yr, the current estimate of *reserves* and *likely resources* of conventional oil will be essentially exhausted over the next 50 years. In addition to conventional oil, we have more than 300 CMO of unconventional petroleum resources (oil shale and tar sands). These unconventional petroleum resources will need to be developed if we are to continue to use oil as a principal energy resource. In the meantime, the challenge will be maintaining the production rate of conventional oil in the face of (a) peaking productivity of many of the world's largest oil fields and (b) lack of new oil field discoveries the size of Ghawar, the largest and most productive field to date. New discoveries today tend to be smaller and increasingly more difficult to access, which basically translates into higher cost for producing oil. Developing these resources requires substantial investments in the range of several hundreds of billions of dollars annually. In 2008, the International Energy Agency projected that a total investment of $6.3 trillion through 2030 would be required to replace the dwindling productivity of the current wells and raise total production to meet the more modest goal of 106 million barrels/day (bpd) in 2030.[2] The 106 million bpd figure corresponds to 1.5 CMO/ yr and is less than earlier estimates, such as that of the Cambridge Energy Research Associates that placed oil demand at 125 million bpd or 1.8 CMO/ yr by 2030.[3]

Natural Gas

The world has about 42 CMO of assured natural gas reserves, and between 34 CMO (likely) and about 66 CMO (speculative) more in conventional

2. International Energy Agency. *World Energy Outlook 2008,* Organisation for Economic Co-operation and Development, Paris, 2008.

3. Peter M. Jackson in "Why the Peak Oil Theory Falls Down: Myths, Legends, and the Future of Oil Resources," Cambridge Energy Research Associates report, Cambridge, MA. 2006.

resource base. Together, this amount could last for about 100 years even with the projected steady increases in consumption rates of natural gas, currently at 0.12 CMO/yr for the United States and 0.65 CMO/yr for the world. Conventional natural gas resources are located principally in the Russian and Middle East regions, and continued reliance on this resource leaves all consuming nations in a politically tenuous position. There are enormous quantities of unconventional natural gas resources in tight sands and gas hydrates (up to 5,000 CMO by some estimates). They are distributed over the globe more widely, but technologies for their cost-effective utilization have yet to be developed.

Coal

This abundant energy source is widely distributed in the world, except in the Middle East region. There are about 120 CMO in reserves, which should be enough to last about 80 years assuming steady growth rates in consumption. Coal consumption in the United States is currently at 0.16 CMO/yr, and for the entire world it is 0.8 CMO/yr. The resource base is estimated at between 400 CMO (likely) and 1,000 CMO (speculative). Coal is the cheapest energy source, and therefore the energy source of choice, particularly—but not exclusively—for the emerging economies of the world that are more concerned with questions of basic existence: food, clothing, and housing for their populations. Coal is also the dirtiest energy source with respect to atmospheric emissions and solid wastes. Whereas measures to combat emissions of pollutants such as nitrogen and sulfur oxides, and toxic metals like mercury and arsenic, can be—and in some instances have been—implemented, controlling the emission of CO_2, the greenhouse gas of chief concern, remains a huge economic challenge.

Capturing CO_2 from the smoke stacks of coal-fired power plants has not been developed at commercial scale. Optimistic studies based on pilot projects suggest that increases of about 25% in the cost of electricity would result from adding CO_2 capture and storage. The problem of storage will not have simple solutions, and there are at present many unknowns that require resolution. If coal is used to provide much of the baseload power for the world's increasing demand for electricity, then we will likely need to increase coal consumption from its current level of 0.8 CMO/yr to more than 2 CMO/yearly by 2060. Producing a CMO of energy would require 5,000 coal-fired plants each rated at 500 MW$_e$, which means that two plants would need to be built each week for the next 50 years.[4] In fact, that is about the rate at which China and India, combined, are currently installing new capacity.

---

4. This and subsequent estimates of the number of plants to be built over 50 years assume that all the plants built are operational at their rated capacity in the 51st year.

Nuclear

Chapter 6 describes how current global production of nuclear power amounts to a little less than 0.2 CMO, and current reserves of uranium (at roughly the yield per ton of current mining practices) are 28 CMO. If we were to increase our annual nuclear energy use to the 1 CMO level, it would appear that we would run out of uranium in a few decades. However, our resource base of uranium could expand 10-fold at a somewhat higher price for uranium, with only a minor impact on the cost of electricity. Availability of fuel could also be extended through reprocessing of spent fuel and severalfold if breeder technologies were adopted. The major challenge with nuclear power is one of public acceptability. Of the non-greenhouse-gas-emitting technologies, nuclear has the most potential for expansion to CMO levels.

A 900-MW nuclear reactor operating at 85% availability, typical for nuclear plants worldwide, would produce 6.7 billion kWh of electricity in a year. At that rate, it would take a battery of 2,283 nuclear plants of 1-GW capacity to produce the electrical energy equivalent of 1 CMO annually. The enormity of the task can be further gauged by considering the rate at which we would need to build such plants. To have 2,283 additional plants for 1 CMO by 2060, we need to be commissioning about 46 per year, or roughly one per week, for 50 years!

Geothermal

Although our current production of energy from geothermal sources amounts to only 0.05 CMO, the global potential for geothermal energy with enhanced energy recovery has been estimated to be as large as 4 CMO/yr, of which about 0.8 CMO/yr would be suitable for electricity production. Once developed, geothermal power facilities have high availability and can be used for base load capacity. We can expect to see a doubling or more of contributions from geothermal energy in the next 10 years, raising this power production to between 0.1 and 0.2 CMO/yr. For geothermal energy to expand beyond that, we will need to develop technologies to extract energy from deep, hot, dry rocks.

Hydroelectric

Hydroelectric power produces about 0.2 CMO/yr globally, and about 1/10th of that in the United States. Because most of the larger rivers of the world have already been dammed, the potential for increasing energy production from this resource is limited. While there is some production from and further interest in small dam and direct run-of-river hydropower production, this has not had, nor is it likely to have, a major impact on the overall status of hydropower. The World Energy Council has estimated that the technically exploitable large-dam hydropower is about 1 CMO/yr, and only a fraction of that—perhaps twice the current level, or 0.4 CMO—would be economical.

Because river flow is not uniform throughout the year, the availability factor for hydroelectric power is about 50%. When completed, the Three Gorges Dam, the world's largest hydroelectric project (22.5 GW), built over 15 years and costing an estimated $30 billion, would generate almost 100 billion kWh annually. If we were to attempt to develop the capability for producing an additional 1 CMO annually from hydropower, we would need to build the equivalent of 153 Three Gorges Dams—about one every four months for the next 50 years!

Wind

The annual global potential for producing energy from wind is more than 40 CMO/yr. Even with practical consideration of land requirements, it should be possible to scale wind power to produce several CMO/yr. The technology for harnessing this income resource is reasonably advanced, and it can produce electricity at 5¢/kWh, which is quite competitive with coal power (~3¢/kWh). The two formidable hurdles facing wind have to do with the intermittency of wind and land requirements. Typically 10 MW of power can be generated from one square mile of land, and availability of wind power is generally around 30%. Under this set of realistic assumptions it would take about three million 2-MW wind turbines, at an estimated cost of more than $8 trillion and covering an area of 580,000 mi^2, to produce 1 CMO/yr. In other words, to have an installed capacity sufficient to generate 1 CMO a year, we would need to commission about 1200 such turbines each week for the next 50 years.

Now, 1200 turbines a week worldwide does not seem an entirely daunting task, and so we can expect wind power to become a substantial contributor in the future. To overcome the intermittency barrier for wind (and solar) facilities and to provide more than a fifth of the total electricity demand, we will need large-scale storage systems or implementation of a smart grid that can integrate power generation from a distribution of fields with compensating quiescent periods. Increased storage capacity could come from advances in battery technologies, massive flywheels, pumped water or air storage systems, or a substantial enhancement of the current grid into a smart system, at a currently unknown but probably large cost. Such a smart system could control the demand side as well as the supply side of power by tapping into the capacity of myriad appliances, including electric cars. The other large hurdle is winning public acceptance for the land required for the wind turbines and associated power lines. The public will have to weigh the benefits of large-scale wind power against its impact on the aesthetics of the environment as well as on farm animals and migratory birds.

Concentrating Solar Power

As discussed in chapter 7, the sun endows us with 23,000 CMO of energy each year. Together with wind, concentrating solar power (CSP) technology is the most ready income-source technology for utility-scale power production. There are a somewhat limited number of locations that can provide sufficient solar intensity on a year-round basis, a fact that can limit solar electric power production. Among the largest CSP plants to date is the 100-MW Andasol project in Spain. There are only a handful of additional such large projects under consideration, and their development is running into other environmental concerns. For example, in California, a state known for its progressive population ready to embrace green energy, two 100-MW projects in the Mojave Desert have been put on hold pending review of their impact on the habitat of a desert turtle. To have an ability to produce 1 CMO from CSP, we will need to build nearly 70,000 projects the size of Andasol around the world at an expense of roughly $14 trillion and cover 27,000 mi^2. In other words, if we opt to follow this path we need to build 27 Andasol projects a week for the next 50 years!

Photovoltaics

Rooftop PV systems, at both residential and commercial buildings, have demonstrated their ability to reduce electric power requirements from conventional sources and thereby reduce greenhouse gas emissions. They are located at the site of energy use and do not require additional power transmission lines. Based on current economics, however, rooftop solar remains the most expensive alternative source of energy. Currently it costs around $20,000 to install a 2-kW system ($10/W), which provides electric power at roughly 35¢/kWh. At this price, the technology can be only a niche player. It appears that government subsidies and rebates, necessary for its growth thus far, will have to continue for PV systems to gain a significant market share. The reason for making this assertion is that the number of installations needed to provide even a fraction of the electricity demand runs into hundreds of millions, which is far more than the number of enthusiastic and wealthy patrons who may be willing to pay the full price. To get 1 CMO/yr of energy from PV systems would require an installed capacity of 8.9 TW. At the current price of $10/W for installed systems, the investment would be $89 trillion. Let's assume that the system price drops five-fold, to $2/W. Even so, the investment cost would be $18 trillion. Taking a commercially available 2.1-kW rooftop system as an example, we can calculate the total number of such rooftop installations for producing 1 CMO. That number calculates out to be 4.2 billion, which means that we need to install PV systems on about a quarter million roofs every day for 50 years!

Naturally, if the installations had larger capacity, the number of systems would be fewer. The largest PV installations to date are only in the few megawatt range, so the total number of installations needed is still staggering. The number and cost of larger, utility-scale PV installations will likely be less, but these systems will probably be located remotely from the energy-using sites and therefore require transmission lines. Because these systems will not have the benefits provided by distributed power generation, the relatively cheaper solar alternative of CSP is likely to be the major player of the two solar electric power technologies for utility-scale installations.

Biomass

At a total global potential of 20 CMO/yr (i.e., the total photosynthetic production of the biosphere), biomass is a resource that must be seriously considered. As a resource that can produce storable fuels and thus reduce our dependence on petroleum for liquid fuels, there is a very compelling case for the development of biomass resources and biofuels from them. That said, all biomass and biofuel processes are not equally scalable. They differ in their impact on the environment or their effect on society through raising the tension between food and fuel needs. There has been a move away from projected use of food crops such as corn and soy to inedible ones such as *Miscanthus* and switchgrass. The processes for producing biofuels from these latter sources are not yet commercial. Even if the crops used to produce biofuels are inedible, they may still compete for the land and water resources used in food production, and thus do not completely alleviate the food-fuel tension.

At an average annual productivity of 5 tons an acre, the area required to produce 1 CMO of biomass would be about 5 million mi^2, or a little less than 1/10th of Earth's land area. Biomass residues from agriculture and forests, and harvesting perennial prairie or savannah growths together with municipal solid waste (mostly paper), can be developed into a sustainable practice, but this resource is limited.

Biofuels produced from algae offer promise. Algae can be grown on marginal lands in saline and brackish waters with annual productivities of 2,000–5,000 gallons of oil per acre. Using an optimistic estimate, the areal requirement for producing 1 CMO of algae-based fuels comes out to 150,000 mi^2, or about 1/30th that from conventional biomass.

What this summary makes abundantly clear is that feasible scale-up of today's set of alternative technologies for producing energy is woefully inadequate to meet even the more moderate estimates of future demands. Pressing forward with their development and acceptance to levels that would make alternative technologies significant contributors to our overall energy needs would be expensive in terms of R&D expenditures and subsidies, which could take resources away from other critical societal needs. The current technologies will likely have undesirable and unintended consequences for the environment that might get overlooked if we rush to implement them on a massive scale. Our income resources for producing energy are extremely demanding of land and have other disadvantages: biomass grown for energy may interfere with the water and land requirements of agriculture; hydropower dislocates large populations and interferes with anadromous fishes such as salmon and shad that must migrate upstream in rivers to lay their eggs; wind turbines can be noisy and unsightly as well as adversely affect migration of certain endangered birds.

Clearly, we will have to make choices. If we want to derive energy from any source, we will have to weigh its benefits against its potential disadvantages. While production in small amounts may be perfectly acceptable, production at the scale that could make a substantial difference in our need for energy might not be. As we weigh these options, we can understand that people from different cultures or economic backgrounds may value things differently. For some, preserving pristine locations is much more important than the energy that could be derived by developing those sites. Others may value the economic development and energy security that developing those locations might afford. The best way to resolve these differences is through an informed debate, and this requires carefully assembled, unbiased information.

Reducing Energy Demand through Efficiency and Conservation

As we saw above, the abilities of various income resources to produce substantial amounts of energy over the next 50 years are quite limited. With these limitations on energy supplies, it is imperative that we pursue ways to *reduce* future energy consumption. Two paths are available. The first is the use of equipment and processes that require less energy for each unit of output, for example, more lumens per watt-hour or higher engine efficiency. Industries that have major energy costs and access to capital funding will respond to price signals quickly. Others will respond less quickly unless rebates or other incentives to buy the more efficient product are offered.

Most of the projected increase in energy demand—perhaps several CMO—is from the developing countries, and that increase far outstrips the savings in energy that could be realized (perhaps 0.5 CMO) if developed nations were to increase their efficiency and cut their own current consumption.

There is an alternative to the search for energy efficiency; it consists of forgoing unnecessary energy uses coupled with remaking our current living environment. This second path, undoubtedly more difficult, is to choose options that require less energy. These range from the simple, such as walking instead of riding or using a renewable bag for transport of groceries, to some remaking of lifestyles, for example, living in apartments rather than single-family homes, to the extremely difficult path of remaking cities and their surroundings so that convenient living would require smaller expenditures of energy.

Policy Considerations

Imposing a Carbon Cost

Substantial new efforts will be required to increase the role of income resources and will very likely increase the cost of energy. To make the income sources somewhat competitive, it would be necessary to increase the cost of energy from fossil sources, through either direct taxes or a cap-and-trade mechanism, which requires the governing body to issue certain emission quotas to different industries and to guess an appropriate level for the maximum permissible emissions (the "cap"). Such a scheme is subject to gamesmanship as companies jockey for emission quotas. In the E.U. countries, a cap-and-trade system has been implemented as a means to curb CO_2 emissions. The U.S. government is thinking of following suit, because proposing new taxes would likely be political suicide under current conditions, even though a direct carbon tax would be more transparent and easier to administer and would give the companies a clear incentive for the objective of reducing demonstrable reductions in carbon emissions. A criticism of the carbon tax is that it would be highly regressive, affecting the low-income users most. This concern can be addressed by using some of the tax revenues as credits to offset increased fuel taxes paid by those in the lower income brackets. The remainder of the revenues could, and *should*, be used to support R&D in energy and clean technologies, as well as public education programs.

Carbon is currently trading on the Carbon Mercantile Exchange Board at around \$20/ton CO_2, and this price is barely making a difference in the consumption of fossil energy. Some economists have argued that in order to have significant impact, the price of carbon has to be increased to more than \$100/ton CO_2. This poses a major problem because in trying to clean up energy, we could transform it into a luxury item that only the wealthy can afford. This effect must be avoided. Energy is important for everyone. Our current technologies for income resources, CO_2 capture from fossil fuel plants, and most energy saving devices add significantly to the cost of energy. We need innovation in all these areas to bring costs down. As a society, we must focus on facilitating the process of developing affordable energy to maintain current supply and add capacity of 4–5 CMO/year over the next 50 years.

Supporting R&D

Research and development activities have historically been important factors in the successes that produced our current society. The required investments could come from the government or the private sector. Our history is replete with examples of both kinds, although in many instances, initial push by military or other government agencies has been instrumental in helping the technologies overcome the hurdle of cost. In the area of energy research, government supported programs have been instrumental in the advances in nuclear power and alternative fuels from coal or biomass. The Manhattan Project, a prime example of government research projects, cost approximately \$2 billion—equivalent to about \$25 billion 2002 dollars—in about two and a half years. It was located at three principal sites, in Tennessee, Washington, and New Mexico, and employed about 130,000 workers of all skills—from theoretical physicists to pipe fitters, university chemists to the unskilled laborers in the construction trades, housemaids, cooks, and janitors. Although the objective of the Manhattan Project was to develop a nuclear weapon, this project laid the foundation for the development of commercial nuclear power and was also incidental to the development of other products, for example, Teflon. Likewise, the more expensive Apollo program, whose mission was to land a man on the Moon and bring him back, required sustained investment of nearly $185 billion over 10 years (in 2002 dollars). This, too, employed people with a wide range of disciplines and skills, and it resulted in many discoveries and innovations, notably in the semiconductor and information technology sector, with broad societal benefits.

Following the oil embargoes of the 1970s, many countries, including the United States, embarked on major programs for developing alternative fuels and other energy sources. The level of funding for these programs covered a longer period than the Manhattan Project, but it was not consistent. When the oil prices dropped, funding for the R&D activities dropped, as well. Several analysts have commented on the decline in the budgetary proportion of energy R&D, as well as on the absolute amount devoted to it. J. J. Dooley analyzed energy R&D budgets of nine OECD countries—Canada, France, Germany, Italy, Japan, the Netherlands, Switzerland, the United Kingdom, and the United States—between 1985 and 1995 and noted that deregulation of electric utilities likely had the unintended consequence of reducing R&D expenditures.[5] In the short run, the reduced R&D spending might be beneficial to the customers in terms of reduced electricity cost, but over the long term the reduced R&D spending would hurt the national energy sectors, their economies, and environmental wellbeing. In an update of that study, P. Runci analyzed the energy R&D budgets since the mid-1990s of 11 industrialized countries: Canada, Denmark, France, Germany, Italy, Japan, the Netherlands, Spain, Sweden, the United Kingdom, and the United States.[6] Among the European countries, energy-related R&D expenditures declined between 95% (United Kingdom) and 55% (France). Most of the decline was in nuclear fission research, but even research on developing income energy sources and conservation also declined. Runci pointed out that even though the United States and Japan have energy R&D expenditures five times the combined outlay of the remaining nine industrialized countries studied, they too have had major cutbacks in energy R&D programs. In Japan and the United States, a doubling of R&D budgets for income sources and conservation has partially countered the decline in other energy programs.

5. J. J. Dooley. "Unintended Consequences: Energy R&D in Deregulated Market," *Energy Policy*, vol. 26, pp. 547–555, 1998.

6. P. Runci. *Energy R&D Investment Patterns in IEA Countries: An Update,* Paper PNWD-3581, Pacific Northwest National Laboratory, 2005.

In a 2007 study, Nemet and Kammen analyzed R&D investments and innovations in the United States.[7] They found that after adjusting for inflation, investment in energy R&D by the U.S. government has declined from about $8 billion (2006 dollars) in the early 1980s to around $4 billion in 2006. The energy R&D investments by industry have also dropped over this period, from $4 billion to less than $2 billion. Thus, the United States as a whole is spending half as much today as it was spending two decades ago on this very important topic. The funding by the government alone has dropped from about 1.0% of the GDP to 0.67%.

The decline in energy R&D has occurred despite calls from many policy analysts and committees for expanding R&D spending and at a time when total federal R&D investments have increased on average 6% per year, and those for health and defense have increased between 10% and 15% per year. The level of funding for energy R&D has been much lower than for many other initiatives that the government has undertaken, such as those for space exploration, homeland security, health research, and strategic defense. In 2006 annual expenditures for health research were about $30 billion, and R&D for defense topped $80 billion, compared to the $2 billion for energy mentioned above.

The question we have to answer is whether we as a society believe energy to be so important for our economy, national security, and the well-being of the people, given that it is receiving so little attention relative to other important goals.

Nemet and Kammen also found a strong positive correlation between R&D expenditures and innovations as reflected in patent filings. Investments in R&D lead to inventions that can reduce costs and improve efficiencies. They concluded that to adequately address the energy crisis, a 5- to 10-fold increase in the energy R&D budget is both feasible and necessary. As we have argued, the world desperately needs innovation in all areas of energy supply and consumption, and increased R&D is an absolute necessity to preparing for additional 4–5 CMO/yr by 2050.

In February 2009 the United States government sharply increased the budget for energy R&D to $15 billion a year for the next ten years. This increased level is in accord with many policy analysts. We can hope that this time there will be consistency in funding.

Investing over the Long Term

Innovations do not happen overnight, so R&D investments have to be consistent over a long time. For example, planning cycles typically span four years

7. G. F. Nemet and D. M. Kammen, *U.S. Energy Research and Development: Declining Investment, Increasing Need, and Feasibility of Expansion,* University of California Postprints, 2007.

in the United States for political reasons, but effective research programs may need funding for development cycles spanning 40 years to cover short-, intermediate-, and long-term objectives. The challenge of providing the world adequate energy for its economic development while also safeguarding the environment is enormous, and none of our current technological solutions can address this dual challenge. Some strategies such as increased efficiency can, and should, be implemented quickly, because they will slow down the growth in energy requirement. There are other technologies, such as nuclear and wind, as well as developing nonconventional fossil resources that could address our needs in the intermediate time frame.

Some of the concepts for advanced combustion engines currently under investigation could offer substantial savings in the intermediate term, as would development of electric and plug-in hybrid cars. Finally, innovations that are currently in a nascent stage, or those that we do not even know of now will require decades before they go through laboratory scrutiny, engineering development, and the maturity needed to play a significant role in the global energy market. To sustain development, our funding of basic R&D has to be for the long haul.

In addition to increasing the energy R&D budget in early 2009, the United States government also passed a $787 billion stimulus package under the American Recovery and Reinvestment Act. Much of this spending is for projects that would lead to employment in the clean energy arena. These are substantial outlays, but there was general support for it because of the economic downturn following the collapse of financial markets and the bankruptcies in the automobile industry. The challenge will be maintaining R&D investment during times when energy prices are low, and public interest wanes, as it did from the mid-1980s to 2004, when oil prices began surging. Communicating the long-range need for energy in readily understandable units, such as the CMO, would be a step in the right direction. It will help us stay focused and give us a framework to choose among the different technologies, because there will always be many other societal needs that also warrant our financial resources. As we choose to invest in various programs, we have to be cognizant of the cost of lost opportunities. Investing in solutions that do not scale to a CMO level will deprive society of investment in other pressing needs, and in essence will constitute a lost opportunity to effect another worthwhile cause, while not really addressing the energy crisis either.

Attitudinal Changes

In preceding chapters we showed that technological advances alone would not be sufficient to deliver us from the energy problem we face. We will need attitudinal changes as well. These changes need to take place in the attitudes of governments, industry, international agencies, nongovernmental organizations, and individuals.

Governments must realize the need to work with other nations to build international partnerships in the development of long-range research and development projects, for example, the ITER fusion research project that the United States has now reluctantly joined. Governments will also need to make sure that the policies they implement have sufficient industry interest and support to be successfull. With the economic crisis that the world has been experiencing since mid-2008, there has been much discussion about the need to create new jobs in conjunction with the decision to provide green energy. Governments are beginning to implement programs to do that, and we applaud those efforts but also caution that in promoting green energy we must guard against inadvertently reducing the total energy supply. The primary purpose of the energy industry is not to provide employment within the industry, but to make a commodity that allows *many* industries to flourish and employ people. Programs for weatherizing homes and building a smart grid are examples of good investments. Raising the cost of electricity by a steep carbon tax could increase employment in the solar and wind sectors, but such a policy could be counterproductive in terms of overall job creation if it stymies growth in other industries in general. Economic models show that an investment of almost $1trillion currently planned under the under the ARRA will more than pay for itself in 10 years, but only if the general economic growth remains robust (more than 2.6%/yr). At lower economic growth rates, the investment will contribute to increasing the national debt.

The discussion of motivations for investments relates to a larger issue, namely, clarity of purpose in public spending. As we have seen, there may be multiple motivations for any policy. Biofuels programs, particularly the corn ethanol program, have been promoted variously as (a) a source of energy, (b) providing energy security, (c) minimizing greenhouse gas emissions, and (d) a means of providing farm subsidies. By putting all the reasons together, we run the risk of turning off the public when it becomes clear, for example, that corn ethanol provides barely 30% extra energy over what was needed to produce it (even after including credits for animal feed coproducts), and it has at best a marginal impact on greenhouse gas emissions. The real benefits were keeping farmers solvent through subsidies and development of an experience base to build on before tackling the technologically more challenging approach of cellulosic ethanol. If these objectives had been clearly articulated in the first place, there might not have been as severe a backlash against corn ethanol.

The energy industry has in the past taken the attitude that rising energy prices will take care of any supply problem by allowing companies to access more difficult to produce resources. In essence they have been saying there will be no great disparity between current and future prices of the more desirable fossil sources. Those companies have been focused on delivering value to their shareholders, while losing sight of their greater responsibility to the public. Can we align the imperatives of generating share-holder value and social responsibility so that both interests are served?

International financing agencies such as the World Bank may look for ways to work with their client countries that are not perceived as either interfering with local government or promoting projects of questionable net environmental and societal benefits, such as large hydropower. The international and intergovernmental organizations already play a very important role in nuclear energy; they will likely also have a major role in establishing and monitoring carbon-trading markets. Nongovernmental organizations have traditionally played the role of blaming the industry for all the evils, and the industry has often blamed NGOs for hampering progress. Meaningful progress on the energy front will require changing this adversarial stance into one that allows a collaborative effort between the industry and these organizations.

While such government activities as the imposition of performance standards for electrical appliances and motor vehicles can help, they will be more effective if the ordinary citizen is persuaded that driving a more efficient vehicle at lower speeds is a positive and significant contribution to reducing our energy consumption. Similarly, operators and tenants of large buildings, including federal, state, and local offices and employees, would need to become convinced that the regulations established benefit the nation as a whole. This is also true of many citizens who build mansions, when a less elaborate house would clearly be adequate, and especially so if these houses are situated in locations that increase the temptation, if not the requirement, for greater use of private transportation. In the absence of willingness to be persuaded to action, we will make less than desirable progress in reducing our national energy demand.

Individuals can have a collective positive impact by making changes in their attitudes. Recognizing the magnitude and consequences of our energy use, we can move to adopting lifestyles that use less energy. Realizing that there are trade-offs with any energy technology, we can resist our temptation to emphasize the negatives, for example, NIMBY (not in my backyard), NOPE (not on planet Earth), or BANANA (build absolutely nothing anywhere near anything).

A Portfolio of Action Items

It is a cliché but nonetheless true to say that there is no single solution to our energy crisis and that a portfolio of solutions will be needed. What we have seen so far is that we do not have enough promising technologies with which to stuff our portfolio to adequately address the projected energy needs of the mid-21st century. Our energy strategy therefore has to involve a phased approach. We must implement some technologies that are mature enough immediately, while we continue to develop those that hold promise of scalability but are too expensive to implement today. With these generalities in mind, we believe that the following short-, intermediate-, and long-term actions are needed. The time frames here refer to when those actions may bear fruit, not to when we need to start work on them, which is *now*.

Short Term (0–10 Years)

Public Education and Energy Literacy

Develop educational resources and systems that enable the general public and decision makers to understand and appreciate the current status and effects of technical and political changes designed to improve energy availability and utilization worldwide. As part of this effort, establish a common vocabulary and reporting system that can be uniformly applied and understood. We have found the use of such units as CMO and GO to be very effective for this purpose.

Increase efforts to provide basic information about energy capacities to members throughout the educational community—elementary grades to high school, adult education, and the college/university level. New training for teachers, media presenters, and their support staff will be necessary before (a) these important communicators can adequately convey the importance of energy, its problems, and potential solutions, and (b) they can discuss the relative impacts and time scales involved in the substantial changes in energy use that the general public is being called upon to implement.

Energy Efficiency

Use the energy we already have more efficiently. Individuals should consider weatherizing their homes and offices. Governmental help in the form of subsidies to encourage replacement of windows, adding more insulation, and installation of solar water heating units should take priority over help for installing PV systems. Efficiency standards for appliances (Energy Star) and buildings (Leadership in Energy and Environmental Design [LEED]) should be periodically revised to reflect best available practices. While these regulations affect the manufacturers, there is also a need for customer awareness. Development of products and services that help customers immediately see their energy use and the impact of their own corrective actions can lead to behavior modification.

Taxes and Incentives

Use appropriate tax penalties and economic incentives to encourage implementation of more energy efficient automobiles and trucks. Raising CAFE standards is a quick way to ease the pressure on oil demand. Higher gasoline taxes like those in much of Europe can also lead to changes in buying and driving patterns that will reduce gasoline usage. The extra money collected should then be devoted to funding of energy-related research.

Efficient Coal Power

Move forward in efforts to exploit our most abundant conventional energy source, coal, by focusing on minimizing the destructive environmental effects of its use. Replacing the current subcritical coal-fired power plants with more efficient supercritical and ultra-supercritical plants, and preferably exploiting combined heat and power output, could save more CO_2 emissions at a lower cost than trying to produce equivalent amounts of energy with wind and solar technologies that are harder to scale.

Income Energy Sources
Support the installation of low-cost wind power and other income source systems in remote areas currently not served by the grid. While these systems will not provide CMO-scale energy, they will greatly improve the standard of living for the local people and defer infrastructure expenses associated with extending power transmission. Secondarily, promote large wind farms and CSP facilities after an open public debate of the issues. Both wind and CSP are scalable technologies and can deliver power economically. They both require substantial land areas and therefore will raise concerns among the local and general population as to their environmental impact. We can hope that through public education and outreach the choices can be laid out and a consensus achieved.

Intermediate Term (Maturing in 10–20 Years)

Unconventional Hydrocarbon Resources
Increase efforts to exploit our unconventional hydrocarbon resources as a temporary measure, necessary during the period before the less environmentally damaging resources become available in sufficient quantities. Oil shale and heavy oils will likely be required to fill the gap between the world's demands for petroleum and its ability to produce oil from conventional sources. Such exploitation may help avoid economic collapse or stagnation but should be done in a way that the environmental and energy costs of such sources are not hidden from the citizenry.

Nuclear Power
Proceed in an orderly manner to increase our use of nuclear power. Increasing the utilization of uranium by reprocessing spent fuel and by developing breeder reactor systems are musts. Reprocessing of spent fuel will reduce the volume of radioactive waste to less than 1/10th of its current level. It will also reduce the time that this waste needs to be safeguarded to about 300 years, and will greatly reduce the risk of exposure from potential leaks to acceptable levels. Reprocessing of fuels and use of breeder technologies also considerably extend the time frame in which nuclear fuel resources could serve us, and these technologies therefore transition nuclear power from a near-term solution to a long-term solution. Development of newer nuclear technologies, with provisions for passive shutdown, should go a long way in allaying fears about plant safety.

We recognize that many people are concerned about long-term disposal of nuclear waste, and more must be done to find and develop acceptable sites and storage procedures. There are also justifiable concerns about misuse of nuclear materials as contaminants and as explosive devices. This threat is real, but probably overblown in magnitude, and must be considered in light of other threats. The best defense is further development and strict implementation of the existing nuclear materials safeguards programs of nuclear nations. With increased support from member nations, the International Atomic Energy Agency could provide more funds and diplomatic access to achieve its goals of monitoring nuclear facilities to ensure safe and secure handling of nuclear products for peaceful purposes.

Biomass

Continue to investigate basic biomass technologies with emphasis on the following:

1. Utilize biomass waste resources, such as corn stover, rice hulls, and forestry residues, as well as grasses and reeds. If economical methods to convert these residues into fuels—by either fermentation or thermal processing—are developed, biomass could displace significant petroleum usage and avoid the adverse effects on food production and greenhouse gas emissions found with the current corn ethanol and biodiesel programs.
2. Develop species like algae that can thrive at the agricultural margins, such as arid or marshy areas not suitable for crop production.
3. Explore options for safely using brackish lakes and oceans as growing areas.

Smart Grid

Consider and redefine the needs for storage and power transmission that will be required for full utilization of photovoltaic and wind power. Meeting these needs will require negotiation and public acceptance with respect to land-use issues, so the public must be informed about them sooner rather than later. Transforming the current grid into a smart one is a major undertaking. Although the smart grid is a prerequisite for large-scale utilization of wind and solar power systems, it also offers wider benefits through enabling many energy efficient applications.

Advanced Personal Transportation

Use high-efficiency engines and light-weight composite materials, currently being designed in laboratories around the world, to build a new fleet of highly efficient cars. We can expect some of these technologies to advance sufficiently over the next decade and begin becoming available in transportation systems thereafter. As the electrical grid gets smarter and cheaper rechargeable batteries become available, we can expect a shift in personal transport from being oil dependent to being powered by electricity. Consider long-range opportunities to reconfigure our current living patterns, that emphasize large dwellings, often located far from both work and life's amenities and necessities, to efficiently organized communities for working and living.

Freight Transportation
Reconfigure or institute modern national freight transportation with long-haul freight rail systems matched with short-haul trucking for local distribution.

Engineered Geothermal
Continue development of technologies for power production from hot dry rocks. The production of geothermal power could be leveraged by advances in the technologies for recovering deeper oil and gas resources.

Long Term (20 or More Years)

Thin-Film Photovoltaics
Continue R&D on thin-film photovoltaics, because they have the potential to dramatically reduce the cost of PV panels. Simultaneously, though, increase their efficiencies for converting sunlight into power needs severalfold. Otherwise, the benefits of low-cost materials could be lost because the increased area of the panels requires a correspondingly larger support structure. To foster a larger role of PV systems, improvements and cost reductions in the supporting structure and ancillary systems, such as for power conditioning and storage, are critically needed—perhaps more so than in the photoactive materials themselves—to bring down the overall cost of PV electricity. Thus, R&D in those relatively infrastructural areas should be included in addition to much needed R&D on novel PV materials.

Biomass for Carbon Capture
Develop plants with increased photosynthetic efficiencies as an objective of synthetic biology. In addition, a rapidly growing plant could be used to reduce CO_2 levels in the atmosphere, and thus offer an inexpensive alternative to the currently expensive carbon-capture and sequestration approaches for reducing atmospheric CO_2 concentration.

Nuclear Fusion
Accelerate the current national and international efforts to demonstrate nuclear fusion, because it would be highly sustainable. Even after a successful demonstration, decades would likely be required before the relatively complicated systems being contemplated will be commercial realities.

This book is a plea for increased and informed dialogue. A democratic society *may* make difficult decisions wisely *if* its citizens are educated about the issues; a democratic society definitely *will not* make wise decisions if its citizens, whose social pressure is needed, are ignorant of the issues. The public at large must get involved in making the critical choices that the society must make in choosing sources for its energy supply. Information is key to such a dialogue.

Information alone, however, is not sufficient. To become a useful motivator, information has to be refracted through the prisms of a range of disciplines. A multifaceted approach with many individuals would also provide a safeguard against our proclivity to favor our own disciplinary preferences. Man, with apologies to Aristotle, is not just a rational animal, he is also a rationalizing animal.[8] There have been and currently are many ad hoc committees and panels convened to discuss and report on the complex nature of the problems involved in energy production, use, and importance to global well-being. We suggest instead that permanently funded and staffed national and international committees be formed. The committee should have the charter to periodically issue reports on the technical, economic, and environmental aspects of energy production and use.

Facing our global energy challenge will require a collaboration of scientists and engineers, with specific knowledge of the technical disciplines involved in energy production and use, along with economists, political scientists, and experts in international relations. It will also require poets, playwrights, and novelists whose view of life is often insightful and valuable, to help translate that information into a vision for the future. And the future begins now.

Continue the discussion at www.oup.com/us/CubicMile.

---

8. For a lucid elaboration of man's rationalizing nature, see Robert A. Heinlein, *Assignment in Eternity,* Baen, Riverdale, NY, 2000 (p. 59).

Index